Lecture Notes
in Control and Information Sciences 316

Editors: M. Thoma · M. Morari

T0191914

R.V. Patel · F. Shadpey

Control of Redundant Robot Manipulators

Theory and Experiments

With 94 Figures

 Springer

Authors

Prof. R.V. Patel
University of Western Ontario
Department of Electrical & Computer Engineering
1151 Richmond Street North
London, Ontario
Canada N6A 5B9

Dr. F. Shadpey
Bombardier Inc.
Canadair Division
1800 Marcel Laurin
St. Laurent, Quebec
Canada H4R 1K2

ISSN 0170-8643

ISBN-10   3-540-25071-9 **Springer Berlin Heidelberg New York**
ISBN-13   978-3-540-25071-5 **Springer Berlin Heidelberg New York**

Library of Congress Control Number: 2005923294

Springer is a part of Springer Science+Business Media

springeronline.com

© Springer-Verlag Berlin Heidelberg 2005
Printed in Germany

Typesetting: Data conversion by author.
Final processing by PTP-Berlin Protago-TEX-Production GmbH, Germany
Cover-Design: design & production GmbH, Heidelberg
Printed on acid-free paper   89/3141/Yu - 5 4 3 2 1 0

To Roshni and Krishna (RVP)

To Lida, Rouzbeh and Avesta (FS)

Preface

This monograph is concerned with the position and force control of redundant robot manipulators from both theoretical and experimental points of view. Although position and force control of robot manipulators has been an area of research interest for over three decades, most of the work done to date has been for non-redundant manipulators. Moreover, while both position control and force control problems have received considerable attention, the techniques for position control are significantly more advanced and more successful than those for force control. There are several reasons for this: First, the effectiveness and reliability of force control depends on good models of the environment stiffness. Second, for stability, servo rates much higher than for position control are needed, especially for contact with stiff environments. Third, techniques that are based on tracking a desired force at the end-effector generally use Cartesian control schemes that are computationally much more intensive and prone to instability in the neighborhood of workspace singularities. The fourth factor is the significantly higher noise that is present in force and torque sensors than in position sensors. While most commercial force sensors are supplied with appropriate filters, the delay introduced by the filters can also affect the accuracy and stability of force control schemes.

A large number of techniques have been developed and used for position control such as Proportion-Derivative (PD) or Proprotional-Integral-Derivative (PID) control, model-based control, e.g., inverse dynamics or computed torque control, adaptive control, robust control, etc. Most of these provide closed-loop stability and good tracking performance subject to various constraints. Several of them can also be shown to have varying degrees of robustness depending on the extent of the effect of unmodeled dynamics or dynamic or kinematic uncertainties.

For force or complaint motion control, there are essentially two main approaches: impedance control and hybrid control. Most techniques currently available are based on one or other of these approaches or a combination of the two, e.g., hybrid-impedance control. Impedance control does

not directly control the force of contact but instead attempts to adjust the
manipulator's impedance (modeled as a mass-spring-damper system) by
appropriate control schemes. For pure position control, the manipulator is
required to have high stiffness and for contact with a stiff environment, the
manipulator's stiffness needs to be low. Hybrid control is based on the
decomposition of the control problem into two: one for the position-controlled subspace and the other for the force-controlled subspace. Hybrid
control works well when the two subspaces are orthgonal to each other.
This decomposition is possible in many practical applications. However, if
the two subspaces are not orthogonal, then contradictory position and force
control requirements in a particular direction may make the closed-loop
system unstable.

From the point of view of experimental results, there have been numer-
ous papers where various position and, to a lesser extent, force control
schemes have been implemented for industrial as well as research manipulators. There have also been a number of attempts made to extend position
and force control schemes for non-redundant manipulators to redundant
manipulators. These extensions are by no means trivial. The main problem
has been to incorporate redundancy resolution within the control scheme to
exploit the extra degree(s) of freedom to meet some secondary task require-
ment(s). With the exception of a couple of papers, these secondary tasks
have been postion based rather than force based. One of the key issues is to
formulate redundancy resolution to address singularity avoidance while satisfying primary as well as secondary tasks. A number of redundancy resolution schemes are available which resolve redundancy at the velocity or
acceleration level. In order to formulate a secondary task involving force
control, it is necessary to resolve redundancy at the acceleration level.
However, this leads to the problem that undesirable or unstable motions can
arise due to self motion when the manipulator's joint velocities are not
included in redundancy resolution.

While considerable work has been done on force and position control
of non-redundant manipulators, the situation for redundant manipulators is
very different. This is probably because of the fact that there are very few
redundant manipulators available commercially and hardly any are used in
industry. The complexity of redundancy resolution and manipulator
dynamics for a manipulator with seven or more degrees of freedom (DOF)
also makes the control problem much more difficult, especially from the
point of view of experimental implementation. Most of the experimental
work done to illustrate algorithms for force and position control of redundant manipulators has been based on planar 3-DOF manipulators. The

notable exceptions to this have been the work done at the Jet Propulsion Laboratory using the 7-DOF Robotics Research Arm and the work presented in this monograph which uses an experimental 7-DOF isotropic manipulator called REDIESTRO.

Acknowledgements

Much of the work described in the monograph was carried out as part of a Strategic Technologies in Automation and Robotics (STEAR) project on Trajectory Planning and Obstacle Avoidance (TPOA) funded by the Canadian Space Agency through a contract with Bombardier Inc. The work was performed in three phases. The phases involved a feasibility study, development of methodologies for TPOA and their verification through extensive simulations, and full-scale experimental implementations on REDIESTRO. Several prespecified experimental strawman tasks were also carried out as part of the verification process. Additional funding, in particular for the design, construction and real-time control of REDI-ESTRO, was provided by the Natural Sciences and Engineering Research Council (NSERC) of Canada through research grants awarded to Professor J. Angeles (McGill University) and Professor R.V. Patel.

The authors would like to acknowledge the help and contributions of several colleagues with whom they have interacted or collaborated on various aspects of the research described in this monograph. In particular, thanks are due to Professor Jorge Angeles, Dr. Farzam Ranjbaran, Dr. Alan Robins, Dr. Claude Tessier, Professor Mehrdad Moallem, Dr. Costas Balafoutis, Dr. Zheng Lin, Dr. Haipeng Xie, and Mr. Iain Bryson. The authors would also like to acknowledge the contributions of Professor Angeles and Dr. Ranjbaran with regard to the REDIESTRO manipulator and the collision avoidance work described in Chapter 3.

R.V. Patel
F. Shadpey

Contents

1 Introduction

The problem of position control of robot manipulators was addressed in the 1970's to develop control schemes capable of controlling a manipulator's motion in its workspace. In the 1980's, extension of robotic applications to new non-conventional areas, such as space, underwater, hazardous environments, and micro-robotics brought new challenges for robotics researchers. The goal was to develop control schemes capable of controlling a robot in performing tasks that required: (1) interaction with its environment; (2) dexterity comparable to that provided by the human arm.

Position control strategies were found to be inadequate in performing tasks that needed interaction with a manipulator's environment. Therefore, developing control strategies capable of regulating interaction forces with the environment became necessary. At the same time, new applications required manipulators to work in *cluttered* and *time-varying* environments. While most non-redundant manipulators possess enough *degrees-of-freedom* (DOFs) to perform their primary task(s), it is known that their limited manipulability results in a reduction in the effective workspace due to mechanical limits on joint articulation and presence of obstacles in the workspace. This motivated researchers to study the role of kinematic redundancy. Redundant manipulators possess more DOFs than those required to perform the primary task(s). These additional DOFs can be used to fulfill user defined additional task(s) such as joint limit avoidance and object collision avoidance. Redundancy has been recognized as a characteristic of major importance for manipulators in space applications. This fact is reflected in the design of Canadarm-2 or the Space Station Remote Manipulator System (SSRMS), a 7-DOF redundant arm, and also the Special-Purpose Dextrous Manipulator (SPDM) [33], also known as Dextre, which consists of two 7-DOF arms.

Finally, imprecise kinematic and dynamic modelling of a manipulator and its environment puts severe restrictions on the performance of control algorithms which are based on exact knowledge of the kinematic and dynamic parameters. This has brought the challenge of developing adaptive/robust control algorithms which enable a manipulator to perform its tasks without exact knowledge of such parameters.

1.1 Objectives of the Monograph

As mentioned in the previous section, various applications of manipulators in space, underwater, and hazardous material handling have led to considerable activity in the following research areas:

- Contact Force Control (CFC) and compliant motion control
- Redundant manipulators and Redundancy Resolution (RR)
- Adaptive and robust control

Position control strategies are inadequate for tasks involving interaction with a compliant environment. Therefore, defining control schemes for tasks which demand extensive contact with the environment (such as assembly, grinding, deburring and surface cleaning) has been the subject of significant research in the last decade. Different control schemes have been proposed: Stiffness control [60], hybrid position-force control [56], impedance control [30], Hybrid Impedance Control (HIC) [1], and robust HIC [40].

Recently, free motion control of kinematically redundant manipulators has been the subject of intensive research. The extra degrees of freedom have been used to satisfy different additional tasks such as obstacle avoidance [6],[14], mechanical joint limit avoidance, optimization of user-defined objective functions, and minimization of joint velocities and acceleration [66]. Redundancy has been recognized as a major characteristic in performing tasks that require dexterity comparable to that of the human arm, e.g., in space applications such as for the SPDM which is intended for use on the International Space Station. However, compliant motion control of redundant manipulators has not attained the maturity level of their non-redundant counterparts. There is not much work that addresses the problem of redundancy resolution in a compliant motion control scheme. Gertz et al. [23], Walker [91] and Lin et al. [39] have used a generalized inertia-weighted inverse of the Jacobian to resolve redundancy in order to reduce impact forces. However, these schemes are single-purpose algorithms, and cannot be used to satisfy additional criteria. An extended impedance control method is discussed in [2] and [51]; the former also includes an HIC scheme.

Adaptive/robust compliant control has also been addressed in recent years [27], [41], and [52]. However, there exists no unique framework for an adaptive/robust compliant motion control scheme for redundant manipulators which enjoys all the desirable characteristics of the methods proposed for each individual area, e.g., existing compliant motion control schemes are either not applicable to redundant manipulators or cannot take full advantage of the redundant degrees of freedom.

The main objective of this monograph is to address the three research areas identified above for redundant manipulators. In this context, existing schemes in each of the three areas are reviewed. Based on the results of this review, a new redundancy resolution scheme at the acceleration level is proposed. The feasibility of this scheme is first studied using simulations on a 3-DOF planar arm. This scheme is then extended to the 3-D workspace of a 7-DOF redundant manipulator. The performance of the extended scheme with respect to collision avoidance for static and moving objects and avoidance of joint limits is studied using both simulations and hardware experiments on REDIESTRO (a REdundant, Dextrous, Isotropically Enhanced, Seven Turning-pair RObot constructed in the Center for Intelligent Machines at McGill University). Based on this redundancy resolution scheme, an Augmented Hybrid Impedance Control (AHIC) scheme is proposed. The AHIC scheme provides a unified framework for combining compliant motion control, redundancy resolution and object avoidance, and adaptive control in a single methodology. The feasibility of the proposed AHIC scheme is studied by computer simulations and experiments on REDIESTRO. The research described in this monograph has addressed the following topics:

- Algorithm development
- Feasibility analysis on a simple redundant 3-DOF planar arm
- Extension of the scheme to the 3D workspace of REDIESTRO
- Stability and trade-off analysis using simulations on a realistic model of the arm and its hardware accessories
- Fine tuning of the control gains in the simulation
- Performing hardware experiments

1.2 Monograph Outline

Chapter 2: *REDUNDANT MANIPULATORS: KINEMATIC ANALYSIS AND REDUNDANCY RESOLUTION*

This chapter introduces the kinematic analysis of redundant manipulators. First, different redundancy resolution schemes are introduced and a comparison between them is performed. Next, the Configuration Control approach at the acceleration level is described. This forms the basis of the redundancy resolution scheme used in the AHIC strategy proposed in Chapter 4. Finally analytical expressions of different additional tasks that can be used by the redundancy resolution module are given and simulation results for a 3-DOF planar arm are presented.

Chapter 3: *COLLISION AVOIDANCE FOR A 7-DOF REDUNDANT MANIPULA-TOR*

This chapter describes the extension of the proposed algorithm for redundancy resolution to the 3D workspace of a 7-DOF manipulator. First, a new primitives-based collision avoidance scheme in 3D space is described. The main focus is on developing the distance calculations and collision detection between the primitives (cylinder and sphere) which are used to model the arm and its environment. Next, the performance of the proposed redundancy resolution scheme is evaluated by kinematic simulation of a 7-DOF arm (REDIESTRO). At this stage, fine tuning of different control variables is performed. The performance of the proposed scheme with respect to joint limit avoidance (JLA), and static and moving object collision avoidance (SOCA, MOCA) is evaluated experimentally using REDIESTRO.

Chapter 4: *CONTACT FORCE AND COMPLIANT MOTION CONTROL*

This chapter begins with a literature review of existing contact force and compliant motion control. Based on this review, a novel compliant and force control scheme Augmented Hybrid Impedance Control (AHIC) is presented. The feasibility of using AHIC to achieve position and force tracking as well as resolving redundancy to perform additional tasks such as JLA, SOCA, MOCA is evaluated by simulation on a 3-DOF planar arm. In addition to the kinematic additional tasks described in Chapter 3, the scheme is capable of incorporating dynamic additional tasks such as multiple-point force control and minimization of joint torques to achieve a desired interaction force with the environment.

Based on the problems encountered (e.g. uncontrolled self-motion and lack of robustness with respect to model uncertainties) during simulations using the AHIC scheme, two modified versions of the original AHIC scheme are proposed. The first scheme aims to achieve self-motion stabilization and also robustness to the manipulator's model uncertainty, while the second scheme introduces an adaptive version of the AHIC controller. Stability and convergence analysis for these two schemes are given in detail. Simulations on a 3-DOF planar arm are carried out to evaluate their performance.

Chapter 5: *AUGMENTED HYBRID IMPEDANCE CONTROL FOR A 7-DOF REDUNDANT MANIPULATOR*

In this chapter, extension of the AHIC scheme to the 3D workspace of REDIESTRO is discussed. Different modules involved in the controller are described. The first step is to extend the algorithm developed in Chapter 4 for the 2D workspace of a 3-DOF planar arm to the 3D workspace of a 7-DOF arm. New issues such as orientation and torque control are discussed. Considering the large amount of computation involved in the controller and the limited processing power available, the next step is to develop control software which is optimized both at the algorithm and code levels. A stability analysis and a trade-off study are performed using a realistic model of the arm and its hardware accessories. Potential sources of problems are identified. These are categorized into two different groups: Kinematic instability due to resolving redundancy at the acceleration level, and lack of robustness with respect to the manipulator's dynamic parameters. These problems are successfully resolved by modification of the AHIC scheme.

Chapter 6: *EXPERIMENTAL RESULTS FOR CONTACT FORCE AND COMPLIANT MOTION CONTROL*

The goal of this Chapter is to demonstrate and evaluate the feasibility and performance of the proposed scheme by hardware demonstrations using REDIESTRO. The first section describes the hardware of the arm (e.g. actuators, sensors, etc.), and the control hardware (VME based controller, I/O interface, etc.). The second section introduces the different software modules involved in the operation, their role, and the communication between different platforms. Before performing the final experimental demonstrations, a detailed analysis is given to provide guidelines in the selection of the desired impedances. A heuristic approach is presented which enables the user to systematically select the impedance parameters based on stability and tracking requirements. Different scenarios are considered and two strawman tasks - surface cleaning and peg-in-the-hole - are performed. The selection is based on the ability to evaluate force and position tracking and also robustness with respect to knowledge of the environment and kinematic errors. These strawman tasks have the essential characteristics of the tasks that SPDM may be required to perform in space - window cleaning and On-Orbit Replaceable Unit (ORU) insertion and removal.

Chapter 7: *CONCLUDING REMARKS*

Based on the proposed algorithms for contact force and compliant motion control of redundant manipulators, general conclusions are drawn concerning the research described in this monograph. Future avenues for research in order to extend the current work are also suggested.

2 Redundant Manipulators: Kinematic Analysis and Redundancy Resolution

2.1 Introduction

Particular attention has been devoted to the study of redundant manipulators in the last 10-15 years. Redundancy has been recognized as a major characteristic in performing tasks that require dexterity comparable to that of the human arm, e.g., in space applications such as in the Special Purpose Dexterous Manipulator (SPDM) of Canadarm-2 designed for the International Space Station. While most non-redundant manipulators possess enough degrees-of-freedom (DOFs) to perform their main task(s), i.e., position and/or orientation tracking, it is known that their limited manipulability results in a reduction in the workspace due to mechanical limits on joint articulation and presence of obstacles in the workspace. This has motivated researchers to study the role of kinematic redundancy. Redundant manipulators possess extra DOFs than those required to perform the main task(s). These additional DOFs can be used to fulfill user-defined additional task(s). The additional task(s) can be represented as kinematic functions. This not only includes the kinematic functions which reflect some desirable kinematic characteristics of the manipulator such as posture control [13], joint limiting [66], and obstacle avoidance [14], [6], but can also be extended to include dynamic measures of performance by defining kinematic functions as the configuration-dependent terms in the manipulator dynamic model, e.g., impact force [39], inertia control [64], etc.

In this chapter, we first give an introduction to the kinematic analysis of redundant manipulators. In the next section, we perform a review of existing methods for redundancy resolution. We also study the performance of different redundancy resolution schemes from the following points of view:

- Robustness with respect to algorithmic and kinematic singularity
- Flexibility with respect to incorporation of different additional tasks

Based on this study, we select one methodology, the "configuration control" approach [63], as the basis for resolving redundancy in the force and compliant motion control schemes that we propose in this monograph for redundant manipulators. We also introduce various choices for the additional tasks and their analytical representations. Simulation results for a 3-DOF planar manipulator are given.

2.2 Kinematic Analysis of Redundant Manipulators

Definition: A manipulator is said to be redundant when the dimension of the task space m is less than the dimension of the joint space n. Let us denote the position and orientation of the end-effector along the axes of interest in a fixed frame by the $(m \times 1)$ vector X, and the joint positions by the $(n \times 1)$ vector q. In the case of a redundant manipulator, $r = n - m$ $(r \geq 1)$ is the *degree of redundancy*. The forward kinematic function is defined as

$$X = f(q) \tag{2.2.1}$$

The differential kinematics are given by

$$\dot{X} = J_e \dot{q} \tag{2.2.2}$$

where $\dot{X}$ is related to the "twist" T_X (vector of linear and angular velocities) of the end-effector by:

$$\dot{X} = H_X T_X \tag{2.2.3}$$

where H_X is a matrix of appropriate dimensions (see [5] for details). The second derivative of X is given by

$$\ddot{X} = J_e \ddot{q} + \dot{J}_e \dot{q} \tag{2.2.4}$$

where J_e is the $(m \times n)$ Jacobian of the end-effector. For a redundant manipulator, equations (2.2.1), (2.2.2) and (2.2.4) represent under-determined systems of equations. J_e can be viewed as a linear transformation mapping from R^n into R^m: The vector $\dot{q} \in R^n$ is mapped into $\dot{X} \in R^m$. Two fundamental subspaces associated with a linear transformation are its null space and its range (Figure 2.1).

The null space, denoted $\aleph(J_e)$, is the subspace of R^n defined by

$$\aleph(J_e) = \{\dot{q} \in R^n | J_e\dot{q} = 0\} \tag{2.2.5}$$

The range denoted $\Re(J_e)$, is a subspace of R^n defined by

$$\Re(J_e) = \{J_e\dot{q} | \dot{q} \in R^n\} \tag{2.2.6}$$

Equation (2.2.5) underlies the mathematical basis for redundant manipulators. For a redundant manipulator, the dimension of $\aleph(J_e)$ is equal to $(n - m')$, where m' is the rank of the matrix J_e. If J_e has full column rank, then the dimension of $\aleph(J_e)$ is equal to the degree of redundancy. The joint velocities belonging to $\aleph(J_e)$, referred to as internal joint motion and denoted by $\dot{q}_\aleph$, can be specified without affecting the task space velocities. Therefore, an infinite number of solutions exists for the inverse kinematics problem. This shows the major advantage of redundant manipulators. Additional constraints can be satisfied while executing the main task specified via positions and orientations of the end-effector. The additional constraints can be incorporated using two different approaches - global and local. Global approaches ([48], [35], and [84]) achieve optimal behavior along the whole trajectory which ensures superior performance over local methods. However, the computational burden of global algorithms makes them unsuitable for real-time sensor-based robot control applications. Hence, we will focus on local approaches.

2.3 Redundancy Resolution

A Cartesian controller generates commands expressed in Cartesian space. In the case of controlling a redundant manipulator, these control inputs should be projected into joint space. Depending on the application requirements and choice of controller, redundancy can be resolved at position, velocity, or acceleration level. In most control schemes, the control input is expressed in form of a reference velocity or acceleration. Therefore, in this section we will focus on the redundancy resolution schemes proposed at velocity or acceleration levels.

2.3.1   Redundancy Resolution at the Velocity Level

Solution of the inverse kinematic problem at the velocity level is of two types - exact and approximate.

2.3.1.1   Exact Solution

For a given $\dot{X}$, a solution $\dot{q}$ is selected which exactly satisfies (2.2.2). Most of the methods are based on the pseudo-inverse of the matrix J_e, denoted by $J_e^\dagger$:

$$\dot{q}_p = J_e^\dagger \dot{X} \tag{2.3.1}$$

The pseudo inverse of J_e can be expressed as

$$J_e^\dagger = \sum_{i=1}^{m'} \frac{1}{\sigma_i} \hat{v}_i \hat{u}_i^T \tag{2.3.2}$$

where the σ_i's, $\hat{v}_i$'s, and $\hat{u}_i$'s are obtained from the singular value decomposition of J_e [25] and the σ_i's are the non-zero singular values of J_e. Equation (2.3.1) represents the general form of a minimum 2-norm solution to the following least-squares problem:

$$min_{\dot{q}}\{\|J_e\dot{q} - \dot{X}\|\} \tag{2.3.3}$$

If J_e has full row rank, then its pseudo inverse is given by:

$$J_e^\dagger = J_e^T(J_e J_e^T)^{-1} \tag{2.3.4}$$

The ability of the pseudo-inverse to provide a meaningful solution in the least-squares sense regardless of whether Equation (2.2.2) is under-specified, square, or over-specified makes it the most attractive technique in redundancy resolution. However, there are major drawbacks associated with this solution. As pointed out in [43], the solution given by (2.3.1) does not guarantee generation of joint motions which avoid singular configurations - configurations in which J_e is no longer full rank. Near singular configurations, the norm of the solution obtained by (2.3.1) becomes very large. This can be seen from a mathematical point of view by (2.3.2), in which the minimum singular value approaches zero ($\sigma_{m'} \to 0$) as a singular configuration is approached, i.e., at a singular configuration, J_e becomes rank deficient. Therefore, as we can see in Figure 2.1, there are some velocities in task space which require large joint rates.

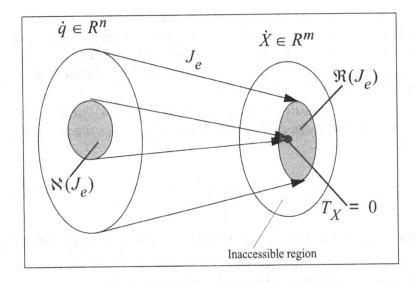

Figure 2.1 Geometric representation of null space and range of J_e

Another problem with the pseudo-inverse approach is that the joint motions generated by this approach do not preserve the repeatability and cyclicity condition, i.e., a closed path in Cartesian space may not result in a closed path in joint space [37]. The final difficulty is that the extra degrees of freedom (when $dim(q) > dim(x)$) are not utilized to satisfy user-defined additional tasks. To overcome this problem, a term denoted $\dot{q}_\aleph$, belonging to the null space of J_e is added to the right hand side of equation (2.3.1) [19].

$$\dot{q} = \dot{q}_p + \dot{q}_\aleph \qquad (2.3.5)$$

Obviously $\dot{q}$ still satisfies (2.2.2). The term $\dot{q}_\aleph$ can be obtained by projection of an arbitrary n-dimensional vector ϑ to the null space of the Jacobian:

$$\dot{q}_\aleph = (I - J_e^\dagger J_e)\vartheta \qquad (2.3.6)$$

where ϑ is selected as follows:

$$\vartheta = \nabla\Phi = \frac{\partial\Phi}{\partial q} = \left[\frac{\partial\Phi}{\partial q_1} \quad \cdots \quad \frac{\partial\Phi}{\partial q_i} \quad \cdots \quad \frac{\partial\Phi}{\partial q_n}\right]^T \qquad (2.3.7)$$

With this choice of the vector ϑ, the solution given by (2.3.5) acts as a gradient optimization method which converges to a local minimum of the cost function. The cost function can be selected to satisfy different objectives, such as torque and acceleration minimization [66], singularity avoidance [47], obstacle avoidance ([14], and [6]).

The other alternative is presented in the so-called extended (augmented) Jacobian methods [21], [61]. The Jacobian of the augmented task is defined by:

$$J_A = \begin{bmatrix} J_e \\ J_c \end{bmatrix} \qquad (2.3.8)$$

where J_A is the extended Jacobian matrix, J_e and J_c being the $(m \times n)$ and $(r \times n)$ Jacobian matrices of the main and additional tasks respectively. The velocity kinematics of the extended task are given by:

$$\dot{Y} = \begin{bmatrix} \dot{X} \\ \dot{Z} \end{bmatrix} = J_A \dot{q} \qquad (2.3.9)$$

where $\dot{X}, \dot{Y}$ and $\dot{Z}$ are the time derivatives of the task vectors of the main, extended and additional tasks, X, Y and Z, respectively. As a result of extending the kinematics at the velocity level, equation (2.3.9) is no longer redundant. Therefore, redundancy resolution is achieved by solving equation (2.3.9) for the joint velocities. However, there are two major drawbacks associated with this method [64]:

(i) The dimension of the additional task should be equal to the degree of redundancy which makes the approach not applicable for a wide class of additional tasks, such as those additional tasks that are not active for all time, e.g., obstacle avoidance in a cluttered environment.

(ii) The other problem is the occurrence of artificial singularities in addition to the main task kinematic singularities. The extended Jacobian J_A becomes rank deficient if either of the matrices J_e or J_c is singular, or there is a conflict between the main and additional tasks (which translates into linear dependence of the rows of J_e and J_c). In practical applications, the singularities of the end-effector are too complicated to determine *a priori*. Furthermore, the singularities of J_c are task dependent which makes them hard to determine analytically. Therefore, the solution of (2.3.9) based on the inverse of the extended Jacobian J_A may result in instability near a singular configuration.

2.3.1.2 Approximate Solution

An alternative approach to dealing with the problem of artificial/kinematic singularities and large joint rates is to solve this problem for an approximate solution. The idea is to replace the exact solution of a linear equation, as in (2.2.2), with a solution which takes into account both the accuracy and the norm of the solution at the same time. This method which was originally referred to as the damped least-squares solution, has been used in different forms for redundancy resolution [92], [47]. The least-squares criterion for solving (2.2.2) is defined as follows:

$$\left\| J_e \dot{q} - \dot{X} \right\|^2 + \lambda^2 \|\dot{q}\|^2 \tag{2.3.10}$$

where λ, the damping or singularity robustness factor, is used to specify the relative importance of the norms of joint rates and the tracking accuracy. This is equivalent to replacing the original equation (2.2.2) by a new augmented system of equations represented by:

$$\begin{bmatrix} J_e \\ \lambda I \end{bmatrix} \dot{q} = \begin{bmatrix} \dot{X} \\ 0 \end{bmatrix} \tag{2.3.11}$$

and finding the least-squares solution for the new system of equations (2.3.11) by solving the following consistent set of equations:

$$(J_e^T J_e + \lambda^2 I)\dot{q} = J_e^T \dot{X} \tag{2.3.12}$$

The least-squares solution is given by:

$$\dot{q}^{(\lambda)} = (J_e^T J_e + \lambda^2 I)^{-1} J_e^T \dot{X} \tag{2.3.13}$$

The practical significance of this solution is that it gives a unique solution which most closely approximates the desired task velocity among all possible joint velocities which do not exceed . The singular value decomposition

$$\|\dot{q}^{(\lambda)}\|$$

(SVD) of the matrix in (2.3.13) is given by:

$$(J_e^T J_e + \lambda^2)^{-1} J_e^T = \sum_{i=1}^{m'} \frac{\sigma_i}{\sigma_i^2 + \lambda^2} \hat{v}_i \hat{u}_i^T \tag{2.3.14}$$

where σ_i's, $\hat{v}_i$'s, and $\hat{u}_i$'s are as in (2.3.2). By comparing the above SVD with that in (2.3.2), we notice a close relationship. Setting $\lambda = 0$, we obtain the pseudo inverse solution from (2.3.14). Moreover, if the singular values are much larger than the damping factor (which is likely to be true far from singularities), then there is little difference between the two solutions, since in this case:

$$\frac{\sigma_i}{\sigma_i^2 + \lambda^2} \approx \frac{1}{\sigma_i} \tag{2.3.15}$$

On the other hand, if the singular values are of the order of λ (or smaller), the damping factor in the denominator tends to reduce the potentially high norm joint rates. In all cases, the norm of joint rates will be bounded by:

$$\|\dot{q}^{(\lambda)}\| \le \frac{1}{2\lambda} \|\dot{X}\| \tag{2.3.16}$$

Figure 2.2 shows the comparison between solutions obtained by the two methods. As we can see, the two problems associated with the pseudo inverse – discontinuity at singular configurations and large solution norms near singularities, are modified in the damped least-squares solution. Based on this, Seraji [63], [66], and Seraji and Colbaugh [65] proposed a general framework for redundancy resolution, referred to as *Configuration Control*.

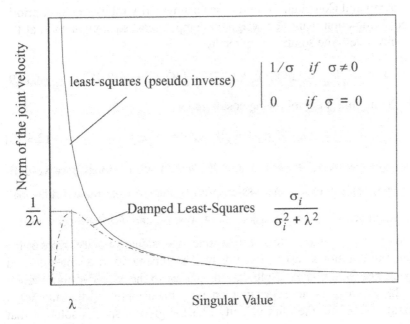

Figure 2.2 Damped versus undamped least-square solution

2.3.1.3 Configuration Control

Under Configuration Control, the $(m \times 1)$ main task vector X is augmented by the $(k \times 1)$ additional task vector Z, and the augmented $[(m + k) \times 1]$ task vector is defined by $Y^T = [X^T, Z^T]^T$. The augmented differential kinematics are given by:

$$\dot{Y} = \begin{bmatrix} \dot{X} \\ \dot{Z} \end{bmatrix} = J_A \dot{q} \qquad (2.3.17)$$

where

$$J_A = \begin{bmatrix} J_e \\ J_c \end{bmatrix} \qquad (2.3.18)$$

is the augmented Jacobian matrix, J_e and J_c being the $(m \times n)$ and $(k \times n)$ Jacobian matrices of the main and additional tasks respectively.

Seraji and Colbaugh [65] proposed a singularity robust and task prioritized formulation using the weighted damped least-squares method at the
velocity level. The solution is given by:

$$\dot{q} = [J_e^T W_e J_e + J_c^T W_c J_c + W_v]^{-1}[J_e^T W_e \dot{X}^d + J_c^T W_c \dot{Z}^d] \qquad (2.3.19)$$

which minimizes the following cost function:

$$\Upsilon = \dot{E}_e^T W_e \dot{E}_e + \dot{E}_c^T W_c \dot{E}_c + \dot{q}^T W_v \dot{q} \qquad (2.3.20)$$

where $W_e(m \times m)$, $W_c(k \times k)$ and $W_v(n \times n)$ are diagonal positive-definite weighting matrices that assign priority between the main, additional,
and singularity robustness tasks. $\dot{E}_e = \dot{X} - \dot{X}^d$ and $\dot{E}_c = \dot{Z} - \dot{Z}^d$ are the
n- and k-dimensional vectors representing the residual velocity errors of the
main and additional tasks respectively. The superscript d denotes desired
trajectories for the tasks. Note that in contrast to the extended formulation
in (2.3.9), there is no restriction on the dimension(s) of the additional
task(s). Therefore, the joint velocity (2.3.19) gives a special solution that
minimizes the joint velocities when $k < r$, i.e., there are not as many active
tasks as the degree-of-redundancy, and the best solution in the least-squares
sense when $k > r$. In all cases the presence of W_v ensures the boundedness
of joint velocities.

2.3.1.4   Configuration Control (Alternatives for Additional Tasks)

Configuration control can serve as a general framework for resolving
redundancy. Any additional task represented as a kinematic function can be
incorporated in this scheme [66]. This not only includes the kinematic functions which reflect some desirable kinematic characteristics of the manipulator such as posture control, joint limiting, and obstacle avoidance, but can
also be extended to include dynamic measures of performance by defining
kinematic functions as the configuration-dependent terms in the manipulator dynamic model, e.g., contact force, inertia control, etc. [64].

In this section, two general approaches for representing additional tasks
are formulated:

(i) *Inequality constraints:* In many applications, the desired additional
task is formulated as a set of inequality constraints $\rho(q) \geq C$, where ρ is a
scalar kinematic function and C is a constant. A kinematic function is

defined as:

$$Z = g(q) = \rho(q) - C; \quad Z^d = 0; \quad \dot{Z}^d = 0; \quad \ddot{Z}^d = 0 \qquad (2.3.21)$$

If $Z > 0$, this task is inactive.

(ii) *Kinematic optimization* of a cost function $\psi(q)$, can be incorporated in configuration control. Additional tasks can be formulated as the following constrained optimization problem:

$$\begin{array}{c} Minimize \quad \psi(q) \\ q \end{array}$$

$$subject \ to \ X - f(q) = 0$$

The solution to this problem can be obtained using Lagrange multipliers. Let the augmented scalar objective function $\psi^\circ(q, \mu)$ be defined as:

$$\psi^\circ(q, \mu) = \psi(q) + \mu^T(X - f(q)) \qquad (2.3.22)$$

where μ is the $(m \times 1)$ vector of Lagrange multipliers. The necessary condition for optimiality can be written as:

$$\frac{\partial}{\partial q}\psi^\circ = 0 \Rightarrow \frac{\partial \psi}{\partial q} = \left(\frac{\partial f}{\partial q}\right)^T \mu = J_e^T \mu \qquad (2.3.23)$$

$$\frac{\partial}{\partial \mu}\psi^\circ = 0 \Rightarrow X = f(q) \qquad (2.3.24)$$

Let N_e be a full rank $(n \times r)$ matrix whose columns span the r-dimensional null space of the Jacobian J_e. The definition of the null space of J_e implies that

$$J_e N_e = 0_{m \times r} \qquad (2.3.25)$$

Pre-multiplying both sides of (2.3.23) by N_e^T yields the optimality condition:

$$N_e^T \frac{\partial \psi}{\partial q} = 0 \qquad (2.3.26)$$

Therefore, the additional task is represented as

$$Z = N_e^T \frac{\partial \psi}{\partial q} \quad and \quad Z^d = 0; \quad \dot{Z}^d = 0; \quad \ddot{Z}^d = 0 \qquad (2.3.27)$$

The Jacobian of the additional task is given by

$$J_c = \frac{\partial Z}{\partial q} = \frac{\partial}{\partial q}\left[N_e^T\left(\frac{\partial \psi}{\partial q}\right)\right] \qquad (2.3.28)$$

2.3.2   Redundancy Resolution at the Acceleration Level

Dynamic control of redundant manipulators in task space, such as the case of compliant control, requires the computation of joint accelerations. Hence, redundancy resolution should be performed at the acceleration level. The second-order differential kinematics are given in (2.2.4). We rewrite the equation as:

$$\ddot{X} - \dot{J}_e \dot{q} = J_e \ddot{q} \qquad (2.3.29)$$

Following the procedure in Section 2.3.1, a similar formulation for $\ddot{q}$ can be obtained to yield exact and approximate solutions. The pseudo-inverse solution is given by:

$$\ddot{q}_P = J_e^\dagger (\ddot{X} - \dot{J}_e \dot{q}) \qquad (2.3.30)$$

where $J_e^\dagger$ is the pseudo inverse of the Jacobian matrix. Equation (2.3.30) represents the general form of a minimum 2-norm solution to the following least-squares problem:

$$min_{\ddot{q}}\{\|J_e\ddot{q} - (\ddot{X} - \dot{J}_e\dot{q})\|\} \qquad (2.3.31)$$

The solutions which are aimed at minimizing the norm of the joint acceleration vector have the shortcoming that they cannot control the joint velocities belonging to the null-space of the end-effector Jacobian or the augmented Jacobian. This may result in internal instability [31]. This problem can be attributed to the instability of the "zero dynamics" of (2.3.29) under a solution of the form (2.3.30) [18]. An example demonstrating this phenomenon is given in Section 4.3.3 . In order to show the source of this problem more clearly, consider a simple kinematic control loop for Cartesian control of a redundant manipulator (Figure 2.3). As we can see in Figure 2.3, the states of the system are q and $\dot{q}$. However, because of the nature of Cartesian control in which the desired trajectory is specified in task space, the terms X and $\dot{X}$ are calculated by applying the nonlinear forward kinematic function to q, and the linear transformation mapping J_e to $\dot{q}$. Let us decompose $\dot{q}$ as follows:

$$\dot{q} = \dot{q}_P + \dot{q}_\aleph \qquad (2.3.32)$$

where

$$\begin{aligned} \dot{q}_\aleph &\in \aleph(J_e) \\ \dot{q}_P &\in \aleph^\perp(J_e) \end{aligned} \qquad (2.3.33)$$

Using the definition of null space, we can write:

$$\dot{X} = J_e\dot{q} = J_e\dot{q}_P + J_e\dot{q}_\aleph = J_e\dot{q}_P + 0 = J_e\dot{q}_P \qquad (2.3.34)$$

This is equivalent to having an open-loop control for the null space component of $\dot{q}$. The question that may be asked is why the pseudoinverse (or configuration control) at the velocity level does not exhibit this phenomenon. The reason is that, the pseudo-inverse solution at the velocity level given by (2.3.1) results in a minimum-norm velocity solution. Therefore, it does not have any null space component. From a mathematical point of view, the pseudo-inverse of J_e is a projector matrix on to $\aleph^\perp(J_e)$. However, the pseudo-inverse solution at the acceleration level results in a minimum-norm acceleration solution which does not guarantee the elimination of the null space component of the velocity.

A solution to this problem was proposed by Hsu et al. [32]. This method requires the symbolic expression of the derivative of the pseudo-inverse of the Jacobian matrix which demands a large amount of computation. A method which combines both computational efficiency with stabilization of internal motion is proposed in Section 5.4.2.1.

2.4   Analytic Expression for Additional Tasks

The general strategies for defining additional tasks – inequality and optimization tasks – were explained in Section 2.3.1.4. In this section, the additional tasks most commonly encountered are formulated analytically under configuration control.

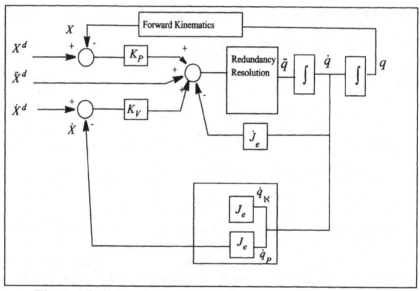

Figure 2.3 Kinematic control loop for a redundant manipulator

2.4.1   Joint Limit Avoidance (JLA)

Joint variables of actual mechanisms are obviously limited by mechanical constraints. In actual implementations, if some joint variables computed by the inverse kinematic module exceed their limits, these joints would be fixed at their extreme values which would restrict movement in certain directions in task space. In this section, we first introduce some relevant terminology, based on which a feasibility analysis of using kinematic redundancy resolution for joint limit avoidance will be presented. Then, we shall use two different approaches for defining algorithms which solve the problem of JLA. The performance of these algorithms will be analyzed by using computer simulations.

2.4.1.1   Definition of Terms and Feasibility Analysis

The *reachable workspace* of a robot manipulator is defined by the geometrical locus of the position and orientation (pose) of the end-effector, $y \in \Re^m$, when the joint variables $q \in \Re^n$, $n \geq m$, range between two extreme values.

$$q_{imin} \leq q_i \leq q_{imax} \qquad\qquad i=1,2,...,n \qquad\qquad (2.4.1)$$

The volume of the reachable workspace is finite, connected and, therefore, is entirely defined by its boundary surface. Obviously on this boundary, some loss of mobility occurs. Therefore the Jacobian matrix becomes rank deficient. The boundary of the reachable workspace can be found numerically by constrained optimization routines, or by applying an inverse kinematics algorithm [62]. As an example, in Figure 2.4, we show the reachable workspace of a two-link manipulator (using an optimization based approach).

In [8] the term "aspects" is used to denote the subspaces of the *accessible volume* in joint space in which the solution of the inverse kinematic function of equation (2.2.1) is unique if $n=m$, or if $n-m$ variables are fixed when $n>m$. The boundaries of the aspects are defined by the singularities of the Jacobian matrix J_e. Therefore, the interior of each aspect is free from singularities. Each aspect in joint space corresponds to a convex subspace of the reachable workspace. In Figure 2.4.a, we show the accessible volume in joint space and its corresponding image in task space (Figure 2.4.b). From these plots, it is obvious that if the desired task trajectory lies inside two different aspects, the inverse kinematics of the manipulator fails to provide a continuous joint trajectory between the initial and the final points. Therefore, this trajectory is not practically realizable without re-configuration of the manipulator at or near the singular configuration. In particular, it is easy to see that for the two-link planar manipulator, with joint limits indicated in Figure 2.4.a and the reachable workspace shown in Figure 2.4.b, we may encounter the following possibilities (Figure 2.5):

- The path AB (the first letter indicates the initial point) is not realizable.

- The path CE via the intermediate point D is not realizable.

- The same path CE via F is realizable.

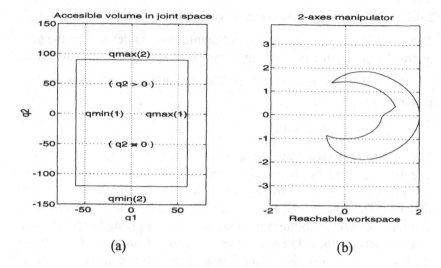

Figure 2.4 Reachable workspace of a 2-DOF manipulator in terms of
a) joint limits, b) reachable workspace

- The path GH with initial joint position $q_2 > 0$ is not realizable.

- The same path GH by the initial configuration $q_2 < 0$ is realizable

Note that by "unrealizable" we mean that there exists no continuous joint trajectory (that can be provided by the inverse kinematics) which starts from the initial configuration and satisfies the task trajectory without violating the joint limits. Thus, for realizing a task comprising motion from an initial pose to a final one, several problems may be considered, and the solutions for some of them may not be achievable by the redundancy resolution module. For instance, task AB is not realizable, but tasks CE and GH can be realized by means of a joint limit avoidance algorithm.

Although the analyzed example is concerned with a non-redundant manipulator, the main concepts are applicable to redundant manipulators under configuration control with the only difference being that, in the redundant case, the augmented task consists of the main and additional tasks which are usually not defined in the same coordinates. Therefore, the geometrical interpretation of the aspects and reachable workspace will, in general, be different in the case of redundant manipulators.

2.4.1.2   Description of the Algorithms

Under the configuration control approach, the criterion of joint limit avoidance should be formulated as a kinematic constraint function. In the following, we present two different approaches for this formulation:

- Using inequality constraints which become active only when one or more of the limits are violated.

- Defining the secondary task as minimization of a desired cost function.

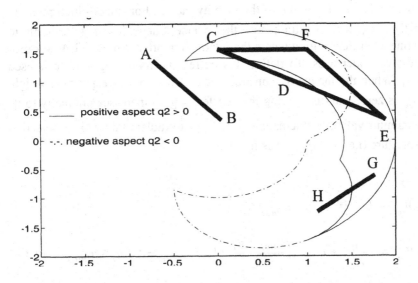

Figure 2.5 Feasibility of different trajectories for a 2-DOF manipulator

2.4.1.3   Approach I: Using Inequality Constraints

In this approach, the basic equations for the JLA algorithm are as follows. The joint limits are presented as a set of inequality constraints. If all the computed values of the joint variables satisfy the inequalities, the redundancy can be used for other tasks. However if one or more of these inequalities are violated, the JLA secondary task should be activated. This task is defined as follows:

$$Z_i = g_i(q) = q_i$$
$$Z_i^d = q_{m_i}$$

(2.4.2)

where q_m replaces either the maximum or the minimum values of the joints for $i=1,2,...,n$, and the corresponding constraint Jacobian J_c is defined by the equation:

$$J_{c_i} = \frac{\partial Z_i}{\partial q} = e_i^T$$

(2.4.3)

where e_i is the *ith* column of the identity matrix. For smooth incorporation of the inequality constraint into the inverse kinematics, it is desirable to define a "buffer" region where the relative importance of the JLA task progressively increases. To define this buffer, the following scheme is used [64]. When the inequality constraint is inactive, the corresponding weight W_{c_i} is zero, and on entering the "buffer" region increases gradually to its maximum value. Mathematically, we can formulate this weight selection procedure (i.e. $q_i \leq q_{imax}$) as follows:

$$
\begin{bmatrix}
W_{c_i} = 0 \quad \text{if} \quad q_i \leq q_{imax} - \tau \\[2ex]
W_{c_i} = \frac{W_0}{4}\left[1 + \cos\left(\pi\left(\frac{q_{imax} - q_i}{\tau}\right)\right)\right] \text{ if } q_{imax} - \tau \leq q_i \leq q_{imax} \quad (2.4.4) \\[2ex]
W_{c_i} = \frac{W_0}{2} \quad \text{if} \quad q_i > q_{imax}
\end{bmatrix}
$$

where W_0 and τ are user-defined constants representing the coefficient for the weight and width of the buffer region respectively.

2.4.1.4   Approach II: Optimization Constraint

The basic idea in the second approach is to define a kinematic objective function which is to be minimized. For joint limit avoidance, the following function has been suggested:

$$\Phi(q) = \sum_{i=1}^{n} \left[\frac{q_i - q_{c_i}}{\Delta q_i} \right]^2 \qquad (2.4.5)$$

where q_c is the center position around which we wish to minimize the movement and Δq is the difference between the maximum and the minimum values of the joints. Then, the redundancy resolution problem is to define a joint trajectory which optimizes equation (2.4.5) subject to the end-effector position.

In Klein [38], it is mentioned that although the quadratic form of equation (2.4.5) is the most used function for this purpose, a better function which reflects the objective of joint limit avoidance has the form:

$$\Phi = max \frac{|q_i - q_{c_i}|}{\Delta q_i} = \left\| \frac{q - q_c}{\Delta q} \right\|_\infty \qquad (2.4.6)$$

However, since the infinity norm is not a differentiable function, he proposed to use some finite order p-norm (p > 2):

$$\Phi = \left\| \frac{q - q_c}{\Delta q} \right\|_p \qquad (2.4.7)$$

For most practical problems, p=6 gives good results. Note that in equation (2.4.7), the different joints have the same importance in the objective function. As an alternative to this formulation, we can introduce a diagonal weighting matrix. The new objective function has the following form:

$$\Phi = \left\| K \left(\frac{q - q_c}{\Delta q} \right) \right\|_p \qquad (2.4.8)$$

where K is an $n \times n$ diagonal matrix. The Jacobian and the desired values for this additional task are calculated as mentioned in and (2.3.28).

2.4.1.5 Performance Evaluation and Comparison

Based on these approaches, two algorithms were implemented. The simulations were carried out on a three-link planar manipulator with link lengths *(0.75m, 0.75m, 0.5m)*, *qmin=[-90 -60 -75]* degrees and *qmax=[45 75 45]*. The reachable workspace and the desired trajectory are shown in Figure 2.6.

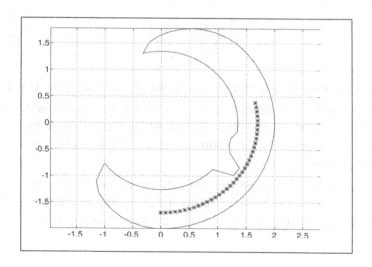

Figure 2.6 Reachable workspace and desired trajectory for a 3-DOF
planar arm

1- Inequality constraint approach: Figure 2.7a shows the joint variables
when the JLA provision was not activated. In this case, the third joint vio-
lates its minimum limit. In the second simulation, the JLA provision based
on the first approach has been used with the nominal selected values
$W_0=100$, $W_v=5$, $W_e=10$, and the buffer region $\tau=5$ (degrees). Figure 2.7b
shows that in this case, the third joint variable does not violate its limit.
Note that by adjusting W_0, the discontinuity of the joint motion resulting
from the nature of the inequality constrained formulation, can be controlled.

   2- Optimization approach: The following simulation used the optimization based JLA ($p=2$). Figure 2.8(a) shows that the third joint variable
enters the buffer region. Figure 2.8(b) shows the results for $p=4$. As we can
see, in this case all joints stay far from their limits. Figure 2.9 shows the
third joint variable for different approaches. As we can see, for this special
case, both methods are successful in following the desired trajectory while
avoiding the joint limits. Obviously, the optimization method ($p=4$) has the
best performance, since, the joint values are kept from approaching the limits. This is in contrast to the inequality approach in which the joints move
freely until coming close to the limits where the JLA becomes active and

prevents the manipulator from exceeding the joint limits. However, the optimization approach is computationally expensive (especially when the number of joints increases) compared to the simple formulation of the inequality constrained approach. Therefore, the inequality constrained approach is preferable for real-time implementations.

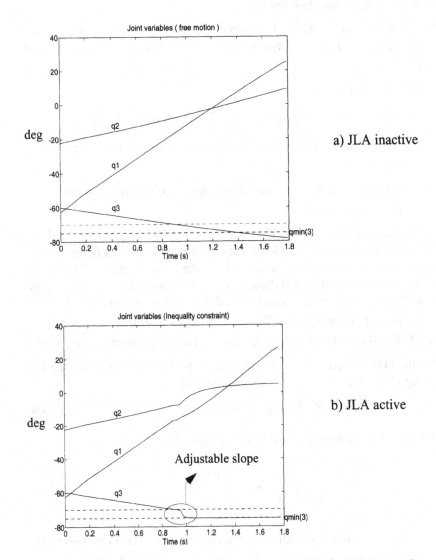

Figure 2.7 Simulation results for JLA using the inequality constraint approach

2.4.2   Static and Moving Obstacle Collision Avoidance

In this section, an outline of an algorithm for the 2-D workspace of a planar arm is given. The extension of the algorithm to a 3-D workspace and simulation results are given in Chapter 3.

2.4.2.1   Algorithm Description

As in the JLA case, Static (and Moving) Obstacle Collision Avoidance is achieved using an inequality constraint. The following steps are followed [14]:

- Distance calculation

- Decision making (if there is a risk of collision for a link)

- Calculation of critical distance - the closest point on the link to the object.

- Utilizing redundancy to inhibit the motion of the critical point towards the object.

For the 2-D workspace, links are modeled by straight lines and the objects are assumed to be circles. Each object is enclosed in a fictitious protection shield (represented by a circle) called the Surface of Influence (SOI). The first step involves distance calculation to find the location of the point X_c (called the critical point) on each link that is nearest to the obstacle by the procedure indicated in Figure 2.10. This algorithm is executed for each link and each obstacle. Then, if any of the critical distances d_{c_i} is less than the SOI, this constraint becomes active. In this case, we define the following kinematic function as the additional task:

$$Z_i = g_i(q, t) = r_O - d_{c_i} \tag{2.4.9}$$

The derivative of the additional task is given below.

$$\dot{Z}_i = \frac{\partial g_i}{\partial q}\dot{q} + \frac{\partial g_i}{\partial t} = -u_i^T\left(\frac{\partial X_{c_i}}{\partial q}\dot{q} - \dot{X}_o\right) \tag{2.4.10}$$

where $\dot{X}_o$ is the time derivative of the object's pose and is related to the object's Cartesian velocity through a linear mapping [5]. The desired values for the active constraints are:

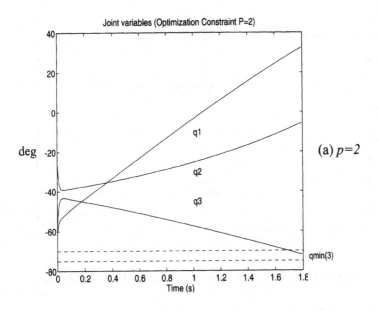

(a) *p=2*

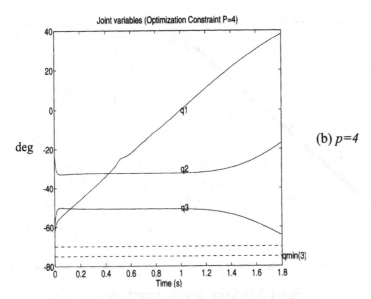

(b) *p=4*

Figure 2.8 Simulation results for JLA using the optimization approach

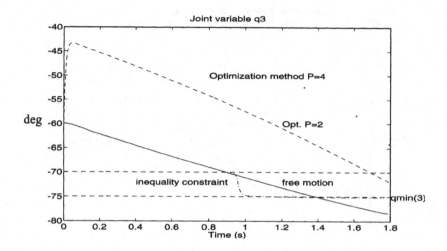

Figure 2.9 Comparison between different JLA approaches

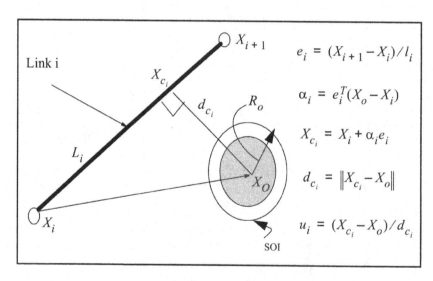

Figure 2.10 Critical distance calculation

$$Z_i^d = \dot{Z}_i^d = \ddot{Z}_i^d = 0 \qquad (2.4.11)$$

Note that we still need to calculate the Jacobian of the active constraints and its derivative. First, an intermediate term is defined as the Jacobian of the critical point, i.e.,

$$J_{X_{c_i}} = \frac{\partial X_{c_i}}{\partial q} \qquad (2.4.12)$$

Then the Jacobian and its derivative are calculated as:

$$J_{c_i} = -u_i^T J_{X_{ci}} \qquad (2.4.13)$$

$$\dot{J}_{c_i} = \frac{\dot{z}_i}{d_{c_i}} u_i^T J_{X_{ci}} + \frac{1}{d_{c_i}}(\dot{X}_{c_i} - \dot{X}_o) J_{X_{ci}} + u_i^T \dot{J}_{X_{ci}} \qquad (2.4.14)$$

2.4.3   Posture Optimization (Task Compatibility)

Compliant motion control and force control are mainly needed for tasks involving heavy interaction with the environment. For this reason, an appealing additional task is to position the arm in a posture which requires minimum torque for a desired force in a certain direction. In this section, a kinematic index for measuring task compatibility is introduced. In section 4.3.2, it is incorporated as an additional task in the Augmented Hybrid Impedance Control (AHIC) scheme.

Similar to the manipulability ellipsoid introduced by Yoshikawa [97], a force ellipsoid can be defined by: $F_e^T(J_e J_e^T)F_e$, where F_e is the environment reaction force. The optimal direction for exerting the force is along the major axis of the force ellipsoid which coincides with the eigenvector of the matrix $J_e J_e^T$ corresponding to its largest eigenvalue (Figure 2.11.a). The force transfer ratio along a certain direction is equal to the distance from the center to the surface of the force ellipsoid along this vector - see Figure 2.11.b where u is the unit vector along the desired direction and α is the force transmission ratio along u. Since αu is a point on the surface of the ellipsoid, it should satisfy the following equation:

$$(\alpha u)^T (J_e J_e^T)(\alpha u) = 1 \qquad (2.4.15)$$

which gives $\alpha = [u^T(J_e J_e^T)u]^{-1/2}$. Hence, Chiu [13] proposed to maximize the following kinematic function (task compatibility index)

$$\phi(q) = \alpha^2 \tag{2.4.16}$$

The desired value and the Jacobian for this additional task can be defined according to the procedure in Section 2.3.1.4 in this chapter. Simulation results are given in Section 4.3.2 .

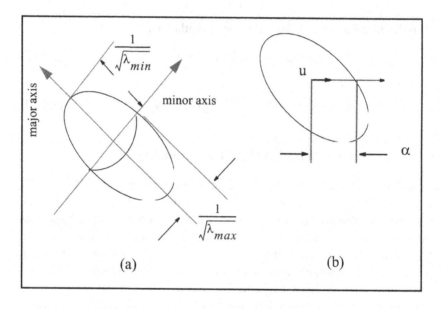

Figure 2.11 a) Force ellipsoid, b) Force transfer ratio in direction u

2.5 Conclusions

In this chapter, the basic issues needed for the analysis of kinematically redundant manipulators were presented. Different redundancy resolution schemes were reviewed. Based on this review, *configuration control* at the acceleration level was found to be the most suitable approach to be used in a force and compliant motion control scheme for redundant manipulators. However, most of the redundancy resolution schemes at the acceleration level suffer from uncontrolled self-motion. In this section, the sources of this problem were presented. Their solutions will be presented in Chapters 4 and 5. The formulation of the additional tasks to be used by the redundancy resolution module were presented in this chapter. Joint limit avoidance which is one of the most useful additional tasks was studied in detail.

The basic formulation of static and moving obstacle collision avoidance task in 2D workspace was presented. We are now in a position to extend the proposed redundancy resolution scheme to the 3D workspace of REDI-ESTRO and evaluate the results by simulation and experiments.

3 Collision Avoidance for a 7-DOF Redundant Manipulator

3.1 Introduction

Collision detection and obstacle avoidance are two features that play an important role in fully or partly autonomous operations of robotic manipulators in cluttered environments. A compact and fast collision-avoidance scheme would be particularly useful for robotic applications in space, underwater, and hazardous environments. Collision avoidance for robot manipulators can be divided into two categories: end-effector level and link level. Much of the work reported to date has dealt with obstacle avoidance as an off-line path planning problem, i.e., find a collision-free path for the end-effector [7], [28], or by mapping the obstacle into joint space, find a collision-free path in joint space [36], [11]. These methods are not applicable to environments with moving objects. Moreover, for non-redundant manipulators, tracking an end-effector trajectory while avoiding collisions with obstacles at the link level, or self-collision avoidance, is often not achievable. Kinematic redundancy has been recognized as a major characteristic for operation of a robot in a cluttered environment [33]. For redundant manipulators, a real-time collision avoidance approach has been developed recently by Seraji and Bon [70] that formulates the problem as a force-control problem so that the task of collision avoidance is solved primarily by augmenting the manipulator control strategy.

To implement a real-time collision-avoidance scheme, three major areas: redundancy resolution, robot and environment modeling, and distance calculation need to be investigated. Obviously, the accuracy with which a robot arm and its environment are modeled is directly related to the real-time control requirements. Greater detail in modeling results in higher complexity when computing the critical distances between an obstacle and the manipulator. Much of this computation can be avoided if the distance measurements are obtained by a proximity sensing system such as the "Sensor Skin" described in [68]. For situations where proximity sensors are not available, a possible solution is to use simple geometric primitives to represent the arm and its environment. Colbaugh *et al.* [14] addressed this problem for a planar manipulator. The obstacles were represented by circles surrounded by a Surface of Influence (SOI), and the links were modeled by straight lines. A redundancy-resolution scheme was proposed to achieve obstacle avoidance. This approach was extended to the 3-D workspace of a 7-DOF manipulator in [75], [71], [24]. In [75] and [71] the manipulator links are represented by spheres and cylinders and the objects by spheres. Although this method is convenient for spherical or bulky objects, it results in major reduction of the workspace when dealing with slender objects. Moreover, the method is not capable of dealing with tasks involving passing through an opening. Glass *et al.* [24] proposed a scheme that considered an application to remote surface inspection. This application requires the robot to pass through circular or rectangular openings for inspection of a space structure, such as the International Space Station. However, they made the restrictive assumption of having an infinite surface with one opening which reduces the workspace of the robot. For instance, this scheme does not permit an "elbow" to back into another opening. Moreover, the arm used in their experiment, the Robotics Research Corporation 7-DOF arm (RRC), is modeled as a series of four straight lines connecting joints one, three, and five. The thickness of the links is considered via a "buffer" region in the openings. This simplified model would not be appropriate for an arm with a more complex geometry such as the one used in the research described in this paper.

A simplified geometrical model for links of industrial manipulators relevant to the study of collisions either with each other or with objects in the workspace is the cylinder. Also, the cylinder is a very appropriate primitive for modelling many objects in the workspace such as rods, mesh structures, openings, etc., without much loss of the available workspace.

In Section 2, we focus on the special cases of sphere-sphere, sphere-cylinder, and cylinder-cylinder collision detection and distance calculations. Considering the importance of cylinder-cylinder collision detection and also its complexity, a novel method of detecting collisions between two cylinders using the notion of dual vectors and angles is presented.

REDIESTRO (Figure 3.1), an isotropic redundant research arm was selected for experimental verification of the collision avoidance system. Its special architecture, resulting from kinematic isotropic design objectives [57], represents a challenge for any collision-avoidance scheme: It has joint offsets, bends in the links, and actuators that are large in relation to the size of the links. It is felt that a successful demonstration of the collision-avoidance scheme on such an arm provides confidence that the system can be developed and applied to other more conventional (i.e., commercial) 7-DOF manipulator designs. Section 3 extends the redundancy-resolution module to the 3D workspace of REDIESTRO. It also describes the incorporation of different additional tasks into the redundancy-resolution module. Simulation results to study the feasibility of the proposed scheme as well as effects of different parameters are given. Section 4 presents the experimental evaluation of the collision- avoidance scheme using REDIESTRO.

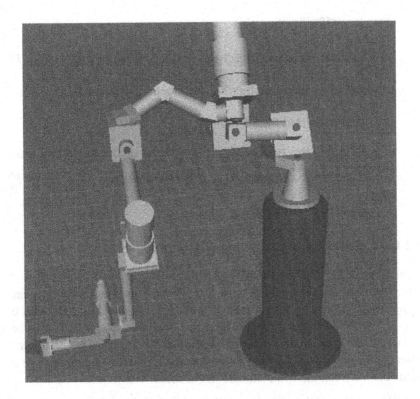

Figure 3.1 Perspective view of REDIESTRO

3.2 Primitive-Based Collision Avoidance

Collision avoidance for stationary and moving objects is achieved by introducing an inequality constraint (see section 2.4.2) as the additional task in the configuration control scheme for redundancy resolution. The idea is to model the links of the manipulator and the objects by primitives such as spheres and cylinders. The major components of the proposed scheme are outlined below:

- **Collision detection/prediction:** For those objects (sub-links) that can potentially collide, determine the critical distance h_{ij}, i.e., the distance from a critical point of the arm to that of the object. The critical points associated with the manipulator and the obstacles are denoted by P_i^c and P_j^c with position vectors p_i^c and p_j^c, respectively.

- **Critical direction detection:** For any pair of critical points P_i^c and P_j^c, determine the critical direction, denoted by u_{ij}, a unit vector directed from P_i^c to P_j^c.

- **Redundancy resolution:** Formulate an additional task and use configuration control to inhibit the motion of the point P_i^c towards P_j^c along u_{ij}.

3.2.1   Cylinder-Cylinder Collision Detection

In order to determine the relative position of two cylinders, first the relative layout of their axes needs to be established. The axes of the cylinders being directed lines in three dimensional space, we resort to the notion of line geometry. Specifically, with the aid of dual unit vectors, (or line vectors), and the dual angles between skew lines, we categorize the relative placement of cylinders and thus determine the possibility and the nature of collisions between the two cylinders in question.

We consider each cylinder to be composed of three parts, the cylindrical surface plus the two circular disks as the top and the bottom of the cylinder. Four points along the axis L_i of each cylinder C_i are of interest (see Figure 3.2), namely, P_i, B_i, T_i, and H_i. The point P_i is any point of reference along the line. The points B_i and T_i with position vectors b_i and t_i, respectively, are the centers of the bottom and top of the cylinder, and H_i is the foot of the common normal of the two lines L_i and L_j on L_i. To avoid ambiguity for the choice of the top and bottom of the cylinder, we can always choose B_i and T_i in such away that the vector $\overrightarrow{B_iT_i}$ points along e_i, with e_i, being a unit vector defining the direction of the cylinder axis (see Figure 3.2). Each of B_i, T_i, and H_i, can alternatively be defined through their line coordinates with respect to the reference point P_i, namely,

$$b_i = p_i + b_i e_i \qquad (3.2.1)$$

$$t_i = p_i + t_i e_i \qquad (3.2.2)$$

$$h_i = p_i + h_i e_i \qquad (3.2.3)$$

It should be noted that for a given cylinder C_i, the scalars b_i and t_i are known and fixed values.

3.2.1.1 Review of Line Geometry and Dual Vectors

A brief review of dual numbers, vectors, and their operations, relevant to our problem is provided in this section. A more detailed discussion can be found in [4], [90], [95]. A line L can be defined via the use of a dual unit vector also called a line vector:

$$\hat{e} = e + \varepsilon m \qquad (3.2.4)$$

where $e^T e = 1$, and $e^T m = 0$, and $\varepsilon \neq 0$ is the dual unity which has the property that $\varepsilon^2 = 0$. Here, e defines the direction of L, while m the moment of L with respect to a self-understood point O, namely,

$$m = p \times e \qquad (3.2.5)$$

with p being the vector directed from O to an arbitrary point P of L. Moreover, e and m are called the primal and dual parts of $\hat{e}$.

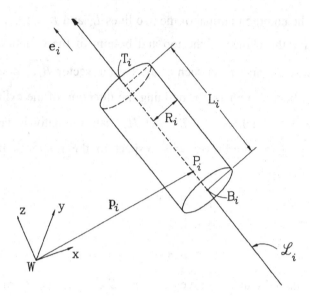

Figure 3.2 Cylinder representation, basic notation.

Now, let L_i and L_j, be two lines. Their *dual angle* is defined as

$$\hat{\upsilon}_{ij} = \upsilon_{ij} + \varepsilon h_{ij} \tag{3.2.6}$$

where υ_{ij} is the *projected angle* between e_i and e_j, and h_{ij} is the *distance* between L_i and L_j. Furthermore,

$$\sin\hat{\upsilon}_{ij} = \sin\upsilon_{ij} + \varepsilon h_{ij}\cos\upsilon_{ij} \tag{3.2.7}$$

$$\cos\hat{\upsilon}_{ij} = \cos\upsilon_{ij} - \varepsilon h_{ij}\sin\upsilon_{ij} \tag{3.2.8}$$

Hence, the dual angle $\hat{\upsilon}_{ij}$ uniquely determines the relative layout of the two lines L_i and L_j in space. Furthermore, the following relations that are in exact analogy with real vectors can be verified:

$$\cos\hat{\upsilon}_{ij} = \hat{e}_i \cdot \hat{e}_j \tag{3.2.9}$$

$$\sin\hat{\upsilon}_{ij} = (\hat{e}_i \times \hat{e}_j) \cdot \hat{n}_{ij} \tag{3.2.10}$$

where $\hat{n}_{ij}$ is the dual vector representing the line N_{ij} that coincides with the common normal of L_i and L_j, and with the same direction as that of the vector from H_i to H_j, namely $\hat{n}_{ij} = n_{ij} + \varepsilon \tilde{n}_{ij}$, where

$$n_{ij} = \frac{h_j - h_i}{\|h_j - h_i\|} \tag{3.2.11}$$

and $\tilde{n}_{ij} = n_{ij} \times h_i = n_{ij} \times h_j$. Hence, equations (3.2.4)–(3.2.11) uniquely determine the dual angle $\hat{\upsilon}_{ij}$ subtended by the two lines. Three different possibilities for the layout of two distinct lines L_i and L_j exist as explained below:

- **(A) Non-Parallel and Non-Intersecting Lines:** $\hat{\upsilon}_{ij}$ is a *proper dual number*, i.e., $\upsilon_{ij} \neq k\pi$, with $k = 0, 1$ and $h_{ij} \neq 0$

- **(B) Intersecting Lines:** $\hat{\upsilon}_{ij}$ is a real number, (its dual part is zero), i.e., $\upsilon_{ij} \neq k\pi$, with $k = 0, 1$ and $h_{ij} = 0$.

- **(C) Parallel Lines:** $\hat{\upsilon}_{ij}$ is a pure dual number, (its primal part is zero), i.e., $\upsilon_{ij} = k\pi$, with $k = 0, 1$ and $h_{ij} \neq 0$.

Now, for two cylinders C_i and C_j to collide, one of the three cases discussed below must occur:

- **(1) Body-Body Collision:** This situation – the most likely one – is shown in Figure 3.3 , where two cylindrical bodies of an object intersect.
- **(2) Base-Body Collision:** The cylindrical body of one cylinder collides with one of the two circular disks of the other cylinder.
- **(3) Base-Base:** One of the circular disks of one cylinder collides with a circular disk of another cylinder.

(A) CYLINDERS WITH NON-PARALLEL AND NON-INTERSECTING AXES

In order to characterize the types of possible collisions for two cylinders whose major axes are represented by L_i and L_j, that are non-parallel and non-intersecting, the following steps are taken:

(a) First we need to determine the location of the points H_i along L_i and H_j along L_j, i.e, the feet of the common normal on the two lines. This can be done by determining the scalars h_i and h_j, as given below:

$$h_i = \frac{(p_i - p_j) \cdot (e_j \cos \upsilon_{ij} - e_i)}{\sin^2 \upsilon_{ij}} \qquad (3.2.12)$$

$$h_j = \frac{(p_j - p_i) \cdot (e_i \cos \upsilon_{ij} - e_j)}{\sin^2 \upsilon_{ij}} \qquad (3.2.13)$$

with $h_i = p_i + h_i e_i$ and $h_j = p_j + h_j e_j$.

(b) Now, if $h_{ij} > (R_i + R_j)$, then collision is not possible.

(c) If $h_{ij} \leq (R_i + R_j)$, then collision is possible, as explained below:

- (A-1)[1] If $b_i \leq h_i \leq t_i$ and $b_j \leq h_j \leq t_j$, then we have a body-body collision and the critical points P_i^c and P_j^c on the axes are H_i and H_j, respectively (Figure 3.3), with the critical direction being n_{ij}.

- (A-2) If only one of the points H_i or H_j lies outside of its corresponding cylinder, then, we may or may not have a collision. However, if the two cylinders collide, then this has to be in the form of a base-body collision only, (Figure 3.4). As an example, in order to determine the critical points and the critical direction, we assume that H_i lies inside C_i with H_j lying outside C_j. The critical point P_j^c of C_j will thus be one of the two points B_j or T_j, whichever lies closer to H_j. Moreover, the critical point P_i^c of the cylinder C_i is the projection of P_j^c on L_i. If p_j^c is the

---

1. *In this notation, the letter indicates the layout of the axes of the two cylinders and the number indicates the type of collision.*

position vector of P_j^c, we have

$$p_i^c = p_i + (\tilde{p}_j^c \cdot e_i)e_i \qquad (3.2.14)$$

where $\tilde{p}_j^c$ is the vector connecting P_j^c to P_i. We thus consider that a collision occurs, whenever the following inequality is satisfied

$$\left\| p_i^c - p_j^c \right\| \le (R_i + R_j)$$

It should be noted that the above inequality gives a conservative prediction of collision between the base and the body of the two cylinders. In this manner, we implicitly assume that the base of the cylinder is not a simple circular disk, but, a fictitious semi-sphere of the same radius. The critical direction u_{ij} for C_i becomes

$$u_{ij} = \frac{p_j^c - p_i^c}{\left\| p_i^c - p_j^c \right\|} \qquad (3.2.15)$$

Case (A-2) above can lead to instability in the redundancy resolution scheme if the two lines are almost parallel. In this special situation, the location of the critical points on the two lines can go through major changes for small changes in the angle υ_{ij} made by them as shown in Figure 3.5. To remedy this "ill-conditioning", we inhibit the motion of two points of the line L_i towards their corresponding projections on L_j whenever the two lines are almost parallel. This is achieved by identifying two critical directions – one for each end of C_i – for the redundancy resolution scheme.

- (A-3) If both H_i and H_j lie outside their corresponding cylinders, then we may have a base-base collision, and the critical points and direction are determined as explained below (Figure 3.6):

  Denote by $\{d_k\}$ the set of distances of B_i and T_i to B_j and T_j, i.e.

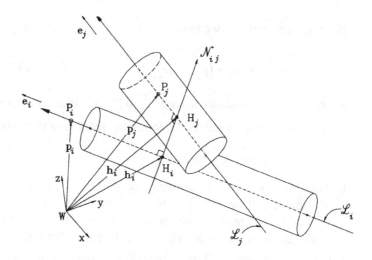

Figure 3.3 (A-1) Body-Body collision (non-parallel and non-intersecting axes)

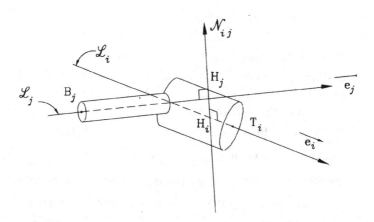

Figure 3.4 (A-2) Base-Body Collision (non-parallel and non-intersecting axes)

$$d_1 = \|b_i - t_j\|, \qquad d_2 = \|b_i - b_j\|$$
$$d_3 = \|t_i - t_j\|, \qquad d_4 = \|t_i - b_j\|$$

$$(3.2.16)$$

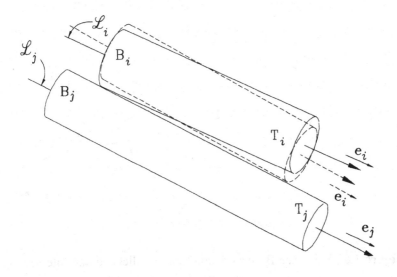

Figure 3.5 Near Parallel axes

and $d_c \equiv min\{d_1, d_2, d_3, d_4\}$, then we have a base-base collision if $d_c \leq (R_i + R_j)$. Once again, the foregoing prediction is conservative, as it assumes two semi-spherical base bodies attached to the ends of the cylinders, rather than the simple circular disks.

(B) CYLINDERS WITH INTERSECTING AXES

In order to characterize a collision between two cylinders with intersecting axes, we first project the end-points B_i and T_i of the cylinder C_i onto the line L_j and denote the projected points by B'_j and T'_j. Conversely, we project the points B_j and T_j of the cylinder C_j onto the line L_i and denote the projected points by B'_i and T'_i. Position vectors of the foregoing four points will take on the form:

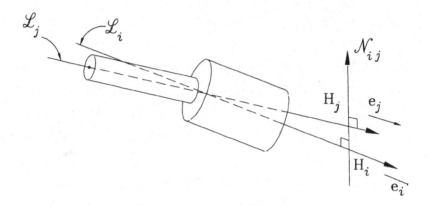

Figure 3.6 (A-3) Base-Base Collision (non-parallel and non-intersecting axes)

$$b'_i = p_i + b'_i e_i, \qquad t'_i = p_i + t'_i e_i$$

$$b'_j = p_j + b'_j e_j, \qquad t'_j = p_j + t'_j e_j$$
(3.2.17)

with

$$b'_i = -(p_i - b_j) \cdot e_i, \qquad t'_i = -(p_i - t_j) \cdot e_i$$

$$b'_j = -(p_j - b_i) \cdot e_j, \qquad t'_j = -(p_j - t_i) \cdot e_j$$
(3.2.18)

- (B-2) If any one of the following four conditions holds, then we have a base-body collision, and the critical direction is a unit vector pointing along a vector joining the corresponding critical points, (Figure 3.7),

$$b_i \le b'_i \le t_i, \quad and, \qquad \|b'_i - b_i\| \le (R_i + R_j)$$

$$b_i \le t'_i \le t_i, \quad and, \qquad \|t'_i - t_i\| \le (R_i + R_j)$$

$$b_j \le b'_j \le t_j, \quad and, \qquad \|b'_j - b_j\| \le (R_i + R_j) \qquad (3.2.19)$$

$$b_j \le t'_j \le t_j, \quad and, \qquad \|t'_j - t_j\| \le (R_i + R_j)$$

- (B-3) If none of the foregoing conditions is satisfied, then we do not have a base-body collision. However, we may have a base-base collision. The procedure for base-base collision detection for a pair of intersecting lines is similar to that of case (A-3) explained earlier (Figure 3.8).

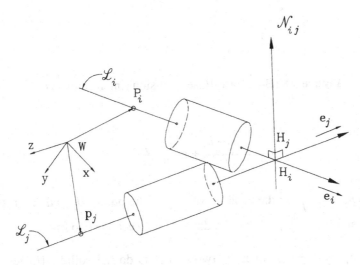

Figure 3.7 (B-2) Base-Body Collision (intersecting axes)

(C) CYLINDERS WITH PARALLEL AXES

For the special case of two parallel lines L_i and L_j for which an infinite number of common normals exist, we resort to a unique definition for one common normal lying closest to the origin [4] – see Figure 3.9. If the line N_{ij} passes through the points H_i and H_j of L_i and L_j (with H_i and H_j being the closest points of the two lines to the origin), then the dual representation of N_{ij} is given as

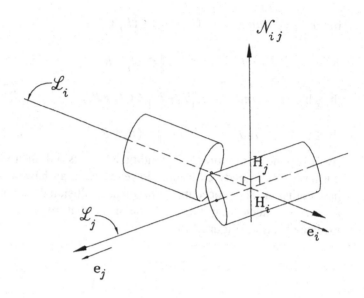

Figure 3.8 (B-3) Base-Base Collision (intersecting axes)

$$\hat{n}_{ij} \ = \ \frac{h_j - h_i}{h_{ij}} + \varepsilon \frac{h_i \times h_j}{h_{ij}} \tag{3.2.20}$$

where h_i and h_j are the position vectors of the points H_i and H_j, respectively, and $h_{ij} \ = \ \|h_j - h_i\|$ is the distance between the two lines.

If $h_{ij} \ge (R_i + R_j)$, then the two cylinders do not collide. However, if $h_{ij} \le (R_i + R_j)$, then, depending on the location of the cylinders along their axes relative to each other, two special cases of body-body (C-1) and base-base (C-3) collisions can occur:

- (C-1) If $h_{ij} \le (R_i + R_j)$, and the projection of either B_i or T_i on L_j is between B_j and T_j, then we have a body-body collision. As in the case of near-parallel axes mentioned in (A-3), to avoid ill-conditioning, we specify two critical directions, one for each end of C_i (Figure 3.5).

- (C-3) If (C-1) is not satisfied, but $h_{ij} \le (R_i + R_j)$, then we obtain the distance between the end points of the two cylinders, as in the (A-3) above (Figure 3.9).

3.2.2 Cylinder-Sphere Collision Detection

This case is simpler than that of cylinder-cylinder collision detection. Figure 3.10 shows the basic layout used for collision detection of the cylinder C_i and the sphere S_j. The notation used for the cylinder is the same as in Section 3.2.1. The sphere S_j is identified by the location of its center P_j and its radius. The first step is to determine if there is a risk of collision. The point H_i on line L_i is determined by projecting the center of the sphere on L_i,

$$h_i = (p_{ij} \cdot e_i)e_i + p_i \qquad (3.2.21)$$

where p_{ij} is defined as $\overrightarrow{P_i P_j}$. The critical distance h_{ij} is given by $h_{ij} = \|p_j - h_i\|$

Now, if $h_{ij} > (R_i + R_j)$, there is no risk of collision. If $h_{ij} \le (R_i + R_j)$, then the following cases can occur:

- If $h_{ij} \le (R_i + R_j)$ and H_i lies inside the cylinder C_i, then the cylinder and the sphere are in collision and the critical points and critical direction are defined by

$$u_{ij} = \frac{p_j - h_i}{\|p_j - h_i\|} \qquad (3.2.22)$$

$$p_i^c = h_i + R_i u_{ij} \qquad (3.2.23)$$

$$p_j^c = p_j - R_j u_{ij} \qquad (3.2.24)$$

- If $h_{ij} \le (R_i + R_j)$ and H_i lies outside the cylinder C_i, then we may or may not have a collision. The critical point on the line L_i is either B_i or T_i depending on which is closer to H_i. Let us

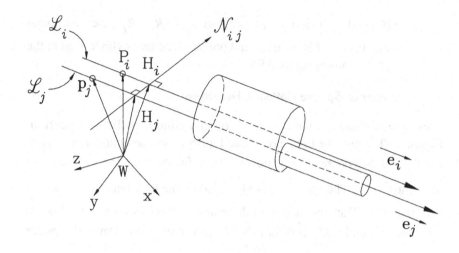

Figure 3.9 (C-3) Base-Base Collision (parallel axes)

assume that B_i is the closer point to H_i. The critical distance h_{ij} is given by $h_{ij} = \|p_j - b_i\|$. Now, if $h_{ij} > (R_i + R_j)$, there is no risk of collision; otherwise, there is a collision and the critical points and direction are calculated by replacing h_i with b_i in equations (3.2.22) through (3.2.24). It has to be mentioned that the foregoing inequality gives a conservative prediction of collision between the sphere and the cylinder. In this manner, we implicitly assume that the base of the cylinder is not a simple circular disk, but a fictitious semi-sphere of the same radius.

3.2.3   Sphere-Sphere Collision Detection

This is the simplest case among the three collision-detection schemes presented. The critical distance h_{ij} is the distance between the centers of the two spheres. If $h_{ij} > (R_i + R_j)$, then there is no risk of collision; otherwise, the two spheres are in collision.

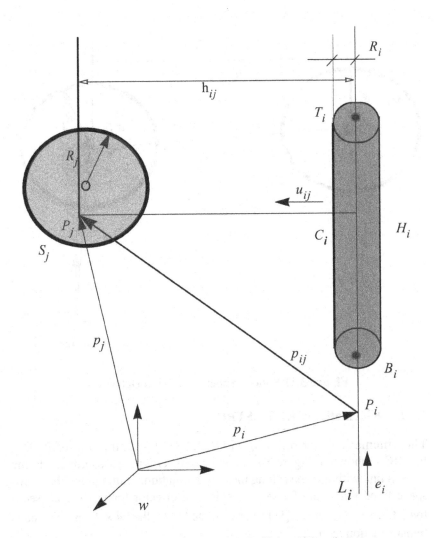

Figure 3.10 Cylinder-Sphere Collision Detection

3.3 Kinematic Simulation for a 7-DOF Redundant Manipulator

In this section, the redundancy-resolution scheme described in Chapter 2 is extended for the general case of a 7-DOF redundant manipulator working in a 3-D workspace and applied to REDIESTRO. The feasibility of the algorithms is illustrated using a kinematic simulation

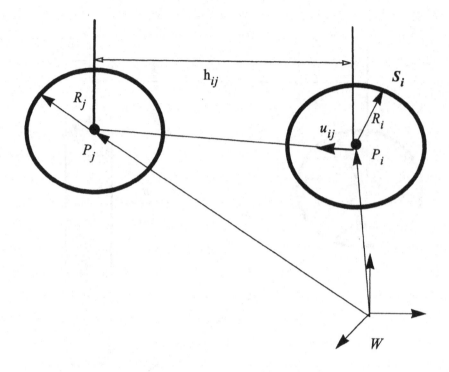

Figure 3.11 Sphere-Sphere Collision Detection

3.3.1   Kinematics of REDIESTRO

The kinematic description of REDIESTRO (a photograph of REDI-
ESTRO is shown in Figure 3.1 is obtained by assigning a coordinate frame
to each link with its z axis along the axis of rotation. Frame {1} is the workspace fixed frame and frame {8} is the end-effector frame. Two consecu-
tive frames {i} and {i+1} are related by the 4×4 homogenous
transformation matrix:

$$
_{i+1}^{i}T = \begin{bmatrix} \cos\theta_i & -\cos\alpha_i\sin\theta_i & \sin\alpha_i\sin\theta_i & a_i\cos\theta_i \\ \sin\theta_i & \cos\theta_i\cos\alpha_i & -\sin\alpha_i\cos\theta_i & a_i\sin\theta_i \\ 0 & \sin\alpha_i & \cos\alpha_i & b_i \\ 0 & 0 & 0 & 1 \end{bmatrix}
$$

$$= \begin{bmatrix} (_{i+1}^{i}R)_{3 \times 3} & (^{i}P_{i+1})_{3 \times 1} \\ 0 & 1 \end{bmatrix} \qquad (i = 1 \to 7) \qquad (3.3.1)$$

where $i = 1 \to 7$; α_i, θ_i, b_i, and a_i are the twist angle, joint angle, offset and link length, respectively. The Denavit-Hartenberg parameters of REDI-ESTRO are given in Table A-1 ( ). The homogenous transformation relating Frame 8 (end-effector frame) to the base frame is given by:

$$_{8}^{1}T = {_{2}^{1}}T{_{3}^{2}}T \ldots \qquad {_{8}^{7}}T \qquad (3.3.2)$$

3.3.2 Main Task Tracking

The main task is described by the pose (position and orientation) of the end-effector, defined by the position vector $(^{1}P_8)_{3 \times 1}$ and the rotation matrix $(_{8}^{1}R)_{3 \times 3}$ of the transformation matrix $_{8}^{1}T$. The pose is thus dimensionally non-homogenous and needs different treatment for the 3-dimensional vector representing the end-effector position from the 3×3 rotation matrix representing orientation. Therefore, the main task is divided into two independent sub-tasks.

3.3.2.1 Position Tracking

The position is described in the workspace-fixed reference frame. Both the desired and the actual position are described in this frame. The ith column of the Jacobian corresponding to the position of the end-effector in frame *{1}* is defined by

$$J_{P_e}(i)_{3 \times 1} = {_{i}^{1}}\hat{Z} \times (^{1}P_{8origin} - {^{1}P_{iorigin}}) \qquad i = 1 \to 7 \qquad (3.3.3)$$

where $_{i}^{1}\hat{Z}$ is the unit vector along the Z axis of joint i, $^{1}P_8$ is the position of the end-effector, and $^{1}P_{iorigin}$ is the position of the origin of the ith frame with respect to frame *{1}*. The position and the velocity errors are given by

$$e_p = {^{1}P_8^d} - {^{1}P_8}, \qquad \dot{e}_p = J_{P_e}\dot{q} - {^{1}\dot{P}_8^d} \qquad (3.3.4)$$

where $\dot{q}$ is the vector of joint velocities, and the superscript d denotes the desired values.

3.3.2.2   Orientation Tracking

The orientation of the end-effector is represented by the 3×3 matrix ^1_8R called the *Direction Cosine* matrix. The *ith* column of the Jacobian matrix, which relates the angular velocity of the end-effector ($^1_e\omega$) to the joint velocity, i.e., $^1_e\omega = J_{O_e}\dot{q}$, can be calculated from the relation

$$J_{O_e}(i) = {}^1_i\hat{Z} \qquad i = 1 \rightarrow 7 \qquad\qquad (3.3.5)$$

The procedure for finding the orientation error and its derivative is more complicated than that for the case of position. In this case, the desired orientation is described by a 3×3 matrix whose columns are unit vectors coincident with the desired x, y, and z axes of the end-effector. The actual orientation of the end-effector is given by the matrix ^1_8R. The orientation error is calculated as follows [42]: $e_O = {}^1K\sin\theta$, where ^1K and θ are the axis and angle of rotation which transform the end-effector frame to the desired orientation. The calculation of the angular velocity error is straightforward:

$$\dot{e}_O = {}^1_e\omega^d - J_{O_e}\dot{q} \qquad\qquad (3.3.6)$$

3.3.2.3   Simulation Results

The performance of redundancy resolution in tracking the main task trajectories is studied here by computer simulation. The integration step size in the following simulations is 10 ms, and the main task consists of tracking the position and orientation trajectories, generated by linear interpolation between the initial and final poses. It should be noted that interpolation of rotations is a much more complex problem than point interpolation in $I\!R^3$. Sophisticated algorithms have been proposed in the literature for this purpose, e.g., see [22], [79], but these are not intended for real-time implementation. For this reason, we use simple linear interpolation for both translation and rotation, which nevertheless leads to satisfactory results. The initial and final poses are specified below:

$$
{}^1P_8^{d-intial} = \begin{bmatrix} 61.8 \\ 231.4 \\ 1127.1 \end{bmatrix}, \quad
{}^1_8R^{d-initial} = \begin{bmatrix} 0.143 & -0.25 & -0.958 \\ -0.93 & 0.30 & -0.22 \\ 0.339 & 0.921 & -0.19 \end{bmatrix}
$$

$$
{}^1P_8^{d-final} = \begin{bmatrix} 500 \\ 500 \\ 1102.3 \end{bmatrix}, \quad
{}^1_8R^{d-final} = \begin{bmatrix} \rho & 0 & \rho \\ 0 & 1 & 0 \\ -\rho & 0 & \rho \end{bmatrix}
$$

where $\rho = \sqrt{2}/2$. The overall redundancy-resolution scheme has not been changed (see Section 2.3.1.3). The only difference consists of splitting the main task into two independent sub-tasks with weighting matrices denoted by W_{P_e} and W_{O_e} corresponding to position and orientation respectively of the end-effector.

The joint velocities are calculated from

$$
\dot{q} = A^{-1}b \tag{3.3.7}
$$

where

$$
A = J_{P_e}^T W_{P_e} J_{P_e} + J_{O_e}^T W_{O_e} J_{O_e} + J_c^T W_c J_c + W_v \tag{3.3.8}
$$

$$
b = J_{P_e}^T W_{P_e} \dot{P}_r + J_{O_e}^T W_{O_e} \Omega_r + J_c^T W_c \dot{Z}_r \tag{3.3.9}
$$

The subscript c refers to the additional task which is not active in the simulation presented in this section. It should be noted in the following simulations that redundancy resolution is implemented in closed-loop. Hence, the reference velocities are given by:

$$
\dot{P}_r = {}^1\dot{P}_8^d + K_{pp}({}^1P_8^d - {}^1P_8) \tag{3.3.10}
$$

$$
\Omega_r = {}^1_e\omega^d + K_{po}e_O \tag{3.3.11}
$$

where K_{pp} and K_{po} are the position and orientation proportional gains respectively. In the first simulation, the sub-task corresponding to tracking the desired orientation is inactive. a and b show the

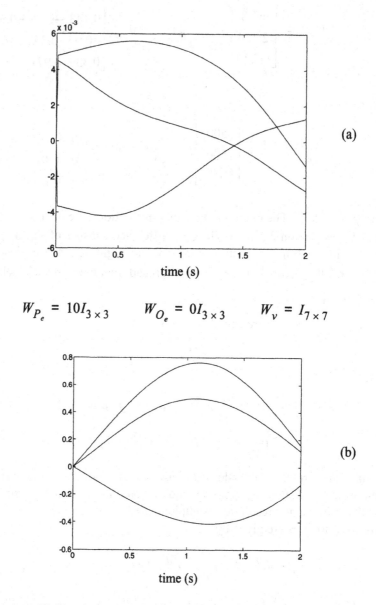

(a)

$$W_{P_e} = 10I_{3 \times 3} \qquad W_{O_e} = 0I_{3 \times 3} \qquad W_v = I_{7 \times 7}$$

(b)

Figure 3.12 Simulation results for position and orientation tracking: (a) position error (m); (b) orientation error (rad)

$$W_{P_e} = 0I_{3 \times 3} \qquad W_{O_e} = 10I_{3 \times 3} \qquad W_v = I_{7 \times 7}$$

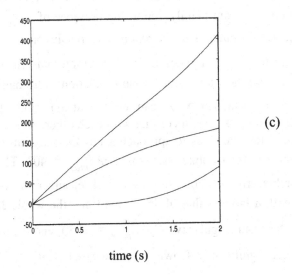

(c)

time (s)

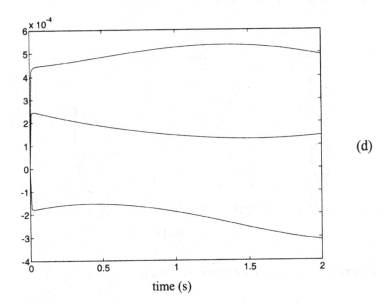

(d)

time (s)

Figure 3.12 (contd.) Simulation results for position and orientation tracking: (c) position error (mm); (d) orientation error (rad)

position and orientation errors. In the second simulation only the orientation sub-task is active, and the results are shown in c,d. In this case, no attempt has been made to follow the position trajectory. The position and orientation errors are mainly due to the presence of W_v in the damped least-squares formulation of the redundancy resolution.

In the following simulations, both position and orientation sub-tasks are active. a-c show the results of the simulation with small W_v (the singularity robustness factor). As we can see in a, at some point, the position and orientation sub-tasks are in conflict with each other. This causes the whole Jacobian of the main task to approach a singular position where the condition number of the Jacobian matrix is $Cond_{max}$ = 403. Therefore, there is considerable error on both sub-tasks.   d, e, and f show the simulation results with a larger value of W_v. This time, the whole Jacobian matrix remains far from singularity ($Cond_{max}$ = 105), and the maximum errors are reduced significantly. However, in the case that W_v = $20I_{7 \times 7}$, there is considerable error at the end of the trajectory. This shows that W_v should be selected as small as possible.

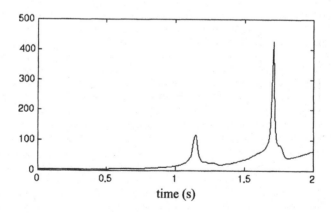

Figure 3.13 (a) Condition number of matrix $\left[J^T_{P_e} , J^T_{O_e} \right]^T$; W_v = $1I_{3 \times 3}$

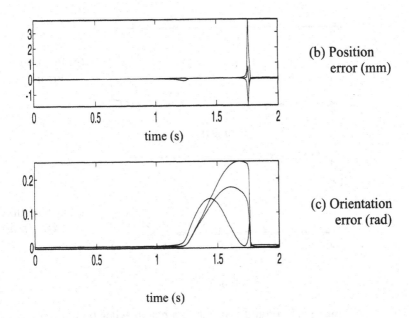

(b) Position
error (mm)

(c) Orientation
error (rad)

time (s)

Figure 3.13 (contd.) Simulation results when both main sub-tasks are active; (a)-(c): $W_v = 1I_{3 \times 3}$

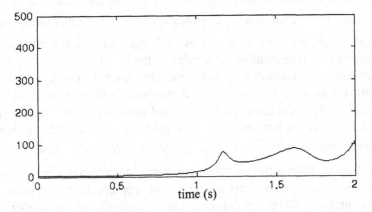

Figure 3.13 (contd.) (d) Condition number of matrix $\left[J^T_{P_e}, J^T_{O_e} \right]^T$;

$$W_v = 1I_{3 \times 3}$$

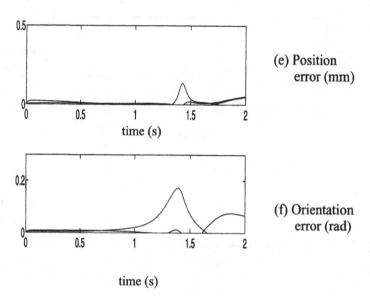

(e) Position
error (mm)

time (s)

(f) Orientation
error (rad)

time (s)

Figure 3.13 (contd.) Simulation results when both main
sub-tasks are active; (d)-(f): $W_v = 20I_{3 \times 3}$

The isotropic design of REDIESTRO reduces the risk of approaching a
singular configuration over a greater part of the workspace. However, this
risk cannot be eliminated completely, and the singularity robustness factor
W_v should either be selected large enough, which introduces errors in the
main task, or should be time-varying, with diagonal entries proportional to
the inverse of the minimum singular value of the "normalized" Jacobian of
the main task. The Jacobian matrix is normalized using the concept of *characteristic length* [85] to resolve the dimensional inhomogeneity in the
matrix due to the presence of positioning and orienting tasks. Figure 3.14
shows the comparison between these two approaches. As one can conclude,
the variable-weight formulation shows better performance because W_v has
small values far from a singular configuration. Hence, variable weights do
not introduce errors on the main task, and grow appropriately near a singular configuration. While the computational complexity of the numerical
implementation of the SVD algorithm for a 7-DOF arm may not be acceptable for real-time control, bounds for the singular values of J can be

found by means of bounds on the real, non-negative eigenvalues of JJ^T. As shown in [86], these bounds can be found quite economically by resorting to the Gerschgorin Theorem [89]

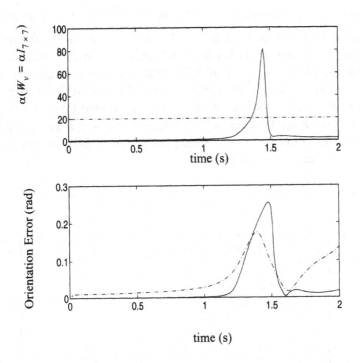

.-.- Fixed W_v, ___ Variable W_v

Figure 3.14 Comparison between the fixed and the time-varying singularity robustness factor

3.3.3   Additional Tasks

The additional tasks incorporated in the redundancy resolution module are as follows: Joint Limit Avoidance (JLA), Stationary and Moving Obstacle Collision Avoidance (SOCA, MOCA) and Self Collision Avoidance (SCA).

3.3.3.1   Joint Limit Avoidance

The JLA algorithm developed in Section 2.4.1.3 is extended here to 3-D without major modifications. In this case, the Jacobian matrix of the JLA corresponding to the *ith* joint is: $J_C = e_i^T$, where e_i is the *ith* column of the matrix $I_{7 \times 7}$. The same weight scheduling scheme is used as that implemented for JLA in Section 2.4.1.3 . In the following simulation, the main task is the same as in Section 3.3.2 with both position tracking and orientation tracking active. Figure 3.15 shows that with JLA inactive, joint 4 has a minimum value equal to 67 degrees. When the JLA is active with minimum 80 degrees for joint 4, this joint is prevented from violating its limit while tracking the main task trajectory. The position and orientation tracking errors converge to small values except for a short transition period when the JLA task becomes active.

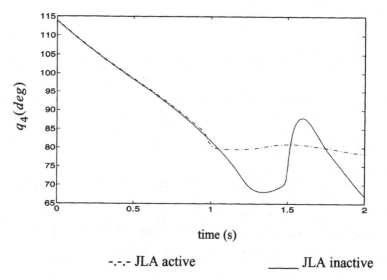

time (s)

-.-.- JLA active   _____ JLA inactive

Figure 3.15 Simulation result for JLA in the 3-D workspace with
$$q_{4_{min}} = 80°$$

3.3.3.2   Stationary and Moving Obstacle Collision Avoidance

A photograph of REDIESTRO, with its actual links and actuators, is shown in Figure 3.1 , while Figure 3.16 depicts the arm with each moving element of the arm enclosed in a cylindrical primitive. The links and the actuator units are modeled by 14 cylinders in total, the fourth link having the maximum number of 4 sub-links. The end-effector and the tool attached to it are enclosed in a sphere.

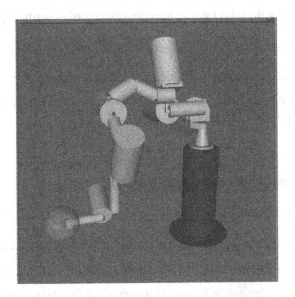

Figure 3.16 REDIESTRO with simplified primitives

The environment is modeled by spherical and cylindrical objects. Each obstacle is enclosed in a cylindrical or a spherical Surface of Influence (SOI). Note that the dimensions of the SOIs are used in distance calculation, collision detection and obstacle avoidance modules rather than the actual dimensions of the obstacles.

<u>Additional task formulation:</u> Let us assume that after performing the distance calculation, the *jth* sub-link of the *ith* link of the manipulator – S_{ij} or C_{ij} depending on the primitive used for modeling – has the risk of collision with the *kth* obstacle (S_k or C_k). The critical point on the sub-link and the obstacle (P_{ij}^c and are P_k^c) and the critical direction ($u_{ij,k}$) are determined by the collision detection algorithm described in Section 3.2 . Now, the additional task z_i for the redundancy-resolution module is defined by:

$$z_{i,k} = h_{ij,k}, \qquad \dot{z}_{ij,k} = -u_{ij,k}^T (J_{ij}^c \dot{q} - \dot{p}_k^c) \qquad (3.3.12)$$

where $h_{ij,k}$ is the critical distance, $J_{ij}^c \dot{q} \equiv \partial \dot{p}_{ij}^c / \partial \dot{q}$ is the Jacobian

matrix mapping the joint rates $\dot{q}$ into the velocity of the critical point P_{ij}^c of

the manipulator, while $\dot{p}_k^c$ is the velocity of the obstacle k. The desired val-

ues for the active constraints (additional tasks) are: $z_i^d = \dot{z}_i^d = 0$. Note

that we still need to calculate the Jacobian of the active constraints. First,

the Jacobian of the critical point is calculated, i.e.,

$$J_{ij}^c = [\bar{J}_{3 \times i} \qquad 0_{3 \times 7 - i}] \qquad (3.3.13)$$

The kth column of the matrix $\bar{J}$ is given by:

$$\bar{J}(k)_{3 \times 1} = \hat{a}_k \times (P_{ij}^c - P_{korigin}) \qquad k = 1 \rightarrow i \qquad (3.3.14)$$

where $\hat{a}_k$ is the unit vector in the direction of rotation of the kth joint,

$P_{korigin}$ is the position vector of the origin of the kth local frame. Note, that

all variables are defined in frame *{1}*. Further, the Jacobian of the additional

task to be used by the redundancy-resolution module is calculated as:

$$J_c = -u_{ij,k}^T J_{ij}^c \qquad (3.3.15)$$

Analysis: The performance of the obstacle avoidance scheme has been

studied by various simulations for different scenarios. As an example, the

simulation results for MOCA are illustrated in Figure 3.17 . In these simu-

lations, the main task consists of keeping the position of the end-effector

constant while avoiding collisions with a moving object. Figure 3.17 shows

the results of the simulations for different constant values of the weighting

matrix corresponding to the collision avoidance task. It should be noted that

when Wc is too small, the object collides with the arm. When W_C is large

enough, no collision occurs, but there is a rapid increase in the joint veloci-

ties which results in a large pulse in joint accelerations (see Figure 3.17). In

a practical implementation, the maximum acceleration of each joint would

be limited and this commanded joint acceleration would result in saturation

of the actuators.

$$R_o = 70 \, mm \text{ and } SOI = 100 \, mm$$

$$\text{- - - } W_c = 0.01 \,, \underline{\quad\quad} W_c = 1 \times 10^{-4} \,,$$

$$\text{-.-.- } W_c = 1 \times 10^{-5} \text{ (collision occurred)}$$

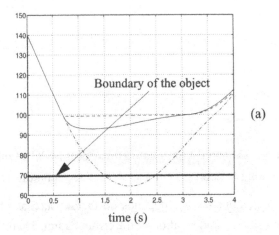

(a)

time (s)

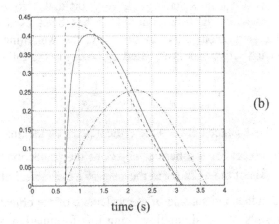

(b)

time (s)

Figure 3.17 Simulation results for MOCA with fixed weighting factors:
(a) Critical distance (mm); (b) 2-norm of joint velocities (rad/s)

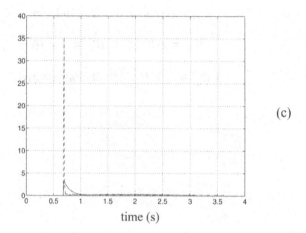

(c)

time (s)

Figure 3.17 (contd.) Simulation results for MOCA with fixed
weighting factors: (c) 2-norm of joint accelerations (rad/s^2)

The optimal value of W_c depends on factors such as object velocity,
end-effector velocity, and location of the critical point. Therefore, from pre-
liminary simulations, it was observed that finding a fixed value which performs well in different situations is very difficult. To overcome this
problem, a time-varying formulation [14] has been used to adjust the
weighting factor automatically. In this way, the weighting factor corresponding to each active task is adjusted according to the following scheme:

$$W_c = k\left(\frac{1}{(d_c - R_O)^2} - \frac{1}{(SOI - R_O)^2}\right) \qquad (3.3.16)$$

where d_c is the distance between the critical point on the link and either the
center of the object for a spherical object or the projection of the critical
point on the axis of the cylinder in the case of a cylindrical object. R_O and

SOI are the radius and surface of the influence of the object respectively.
shows the results of the simulation using this formulation, which for the
case of k = 0.01, shows successful operation of MOCA, with minimum
acceleration.

$- - - k = 100,$ ___ $k = 1,$ $-.-.- k = 0.01$ (obstacle's radius $= 70$ mm and SOI $= 100$ mm)

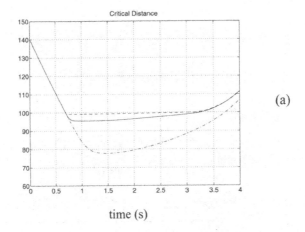

time (s)

(a)

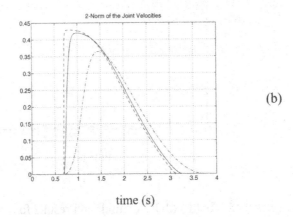

time (s)

(b)

Figure 3.18 MOCA simulation results for time-varying weight factors:
(a) critical distance (mm); (b) 2-norm of joint velocities (rad/s)

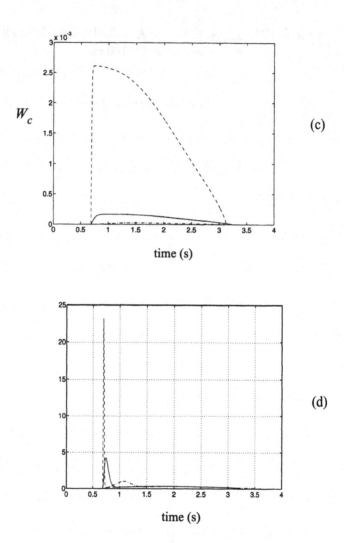

Figure 3.18 (contd.) MOCA simulation results for time-varying weight factors: (c) W_c; (d) 2-norm of joint accelerations (rad/s^2)

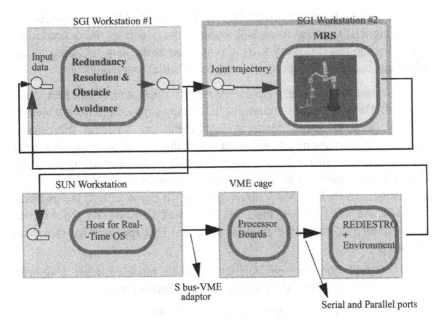

Figure 3.19 General block diagram for the hardware demonstration

3.4 Experimental Evaluation using a 7-DOF Redundant Manipulator

The main objective of these experiments is to demonstrate the capability of the redundancy resolution module in performing the main tasks (position and orientation tracking) while using the extra degrees-of-freedom to fulfill additional tasks (obstacle and joint limit avoidance) for REDIESTRO. The general block diagram of the different modules involved in the hardware experiment is shown in Figure 3.19 .

The three major modules are:

- The redundancy resolution module (RR)
- The robot and its associated control hardware and software
- The robot animation software: Multi-Robot Simulation (MRS) system [9], [10], [77].

In order to distinguish between the performance of the robot controller and the redundancy-resolution scheme, two separate control loops are implemented, one at the Cartesian space level (including the RR) and the other at the low-level joint controller. In this way, the kinematic simulation (including RR) running on an SGI workstation, generates the desired joint trajectory and this trajectory is then transferred as the joint set points to the VME-bus based controller to drive the robot's PID joint controller.

An obstacle-avoidance system essentially deals with a complex environment. There are many limitations in creating (modeling) a robot's environment such as space, material, equipment and financial limitations. Creating a time-varying environment (as in the case of moving obstacles) can be even more difficult. One solution to this problem is online transmission of a robot configuration to a workstation running a graphics visualization of the arm (MRS). MRS serves as a virtual environment; the graphics model of the robot mirrors the exact motion of the arm, and the environment can be modeled in the graphics program. This approach has two main advantages:

- Any complex environment can be modeled with a desired precision (including a time-varying environment)
- The risk of damage to the robot is reduced.

3.4.1   Hardware Demonstration

Three different scenarios were selected to verify the performance of the obstacle-avoidance based redundancy-resolution scheme in executing the following tasks: Position tracking, orientation tracking, stationary and moving obstacle collision avoidance, joint limit, and self-collision avoidance. In each of these scenarios, one or multiple features were active at different instants of execution. The sequence of steps undertaken in each case is as follows:

1. Generate the joint trajectory with the redundancy resolution and obstacle avoidance simulation.

2. Verify the result using MRS (e.g., are the obstacles avoided?).

3. Adjust parameters and repeat step 2 if necessary.

4. Position the stationary obstacles in the workspace.

5. Use the command trajectory to run the robot.

6. Record the joint history for further analysis

For demonstration purposes, the stationary obstacles were built using styrofoam and accurately positioned in the workspace. However, the moving object used in the second scenario was not constructed, instead, the performance of the collision avoidance algorithm was observed using the virtual models of the arm and the object in MRS.

3.4.2 Case 1: Collision Avoidance with Stationary Spherical Objects

In this scenario, the end-effector was commanded to move from its initial position to a final desired position: There were two stationary objects to be avoided in the workspace. The orientation tracking task was not activated in this scenario; the orientation of the end-effector was not controlled. As an example, the plots of the commanded and actual joint values and rates for the first joint are given in Figure 3.20 The set-point command trajectory leads the actual joint trajectory by ~ 0.1 second which is a typical delay of a PID controller (Figure 3.20 a). Figure 3.20 b and c show the desired and actual rates respectively. One can see that the actual rates follow adequately the joint set-point command, except when the joint motion is dominated by stiction. The stiction effects also explain the position error at the end of the trajectory. Note that the PID controller only uses the rate information (obtained by numerically differentiating the measured joint angles) to provide damping. The oscillations shown in the PID rates are probably due to underdamped tuning of the PID parameters and noise due to numerical differentiation.

Figure 3.21 shows the snapshots of the arm motion. We can see that without activating the obstacle avoidance feature (left sequence), the position trajectory is followed perfectly, but, there are several collisions with the obstacles. Figure 3.21 (right sequence) shows the successful operation of position tracking and obstacle avoidance (visualization of the hardware experiment). This scenario demonstrates the capability of the redundancy-resolution module in performing position tracking and avoiding collisions with obstacles.

3.4.3 Case 2: Collision Avoidance with a Moving Spherical Object

In the second scenario, the end-effector was commanded to keep its initial position while the orientation was changed. There was also a moving object to be avoided. In order to satisfy the main task, six DOFs are required, leaving one DOF for additional tasks. Figure 3.22 shows the actual joint angles for joints 2 and 3. The joints initially start moving to realize the commanded change of orientation, but this direction is reversed

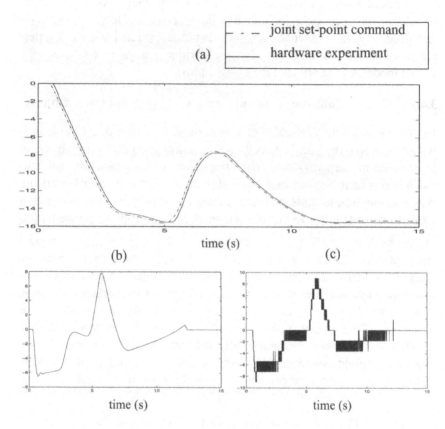

Figure 3.20 Case 1: a) Joint 1 (deg); b) derivative of the joint set-point command (deg/s); c) derivative of joint trajectory in hardware experiment (deg/s).

for joint 2, at 0.9 second, when the arm starts to take evasive action to prevent a collision. The joint-2 angle rapidly increases to a peak value of ~ 30 degrees at 2 seconds. At 2.4 seconds, joint-2 quickly changes its direction to respect the imposed joint limit (software limit to prevent self-collision) of 35°. It should be noted that there are more active additional tasks than the available degrees of redundancy. However, task-prioritized formulation of redundancy resolution is capable of handling these difficult situations and leads only to a graceful performance degradation for the less prioritized tasks (in this case position and orientation tracking).

Figure 3.23 left sequence (simulation results), shows that without any obstacle avoidance, joint-limit avoidance, and self-collision avoidance provisions, only the main task consisting of position and orientation tracking can be successfully executed. However, there are multiple collisions with objects and self-collision with the base. The right sequence of Figure 3.23 shows that by activating different modulesboth the main and additional tasks can be performed simultaneously (visualization of the hardware experiment).

3.4.4   Case 3: Passing Through a Triangular Opening

The environment was modeled by three cylindrical objects forming a triangular opening. The end-effector trajectory was defined as a straight line passing through this opening. Each obstacle is enclosed in a cylindrical SOI. The left column in Figure 3.24 (a–g) shows the motion (simulation results) of the arm when the obstacle-avoidance module is not activated. As can be seen, the end-effector follows the desired trajectory; however, there are multiple collisions between the links or the actuators and the obstacles. By activating the obstacle-avoidance module, both the end-effector trajectory following and obstacle avoidance were achieved, as can be seen in the right column of Figure 3.24 (h–k) – visualization of the hardware experiment.

3.5   Conclusions

In this chapter, the extension of the redundancy-resolution and obstacle-avoidance module to the 3D workspace of REDIESTRO was addressed. The obstacle-avoidance algorithm was modified to consider 3-D objects. A primitives-based collision-avoidance scheme was described. This scheme is general, and provides realism, efficiency of computation, and economy in the use of the amount of free space around a redundant manipulator. Different possible cases of collisions were considered. In particular, cylinder-cylinder collision avoidance which represents a complex case for a collision-detection scheme was formalized using the notion of dual vectors and angles.

Before performing the hardware experiments using REDIESTRO to evaluate the performance of the redundancy-resolution and obstacle-avoidance modules, extensive simulations were performed using the kinematic model of REDIESTRO. These simulations were aimed at a study of the following issues:

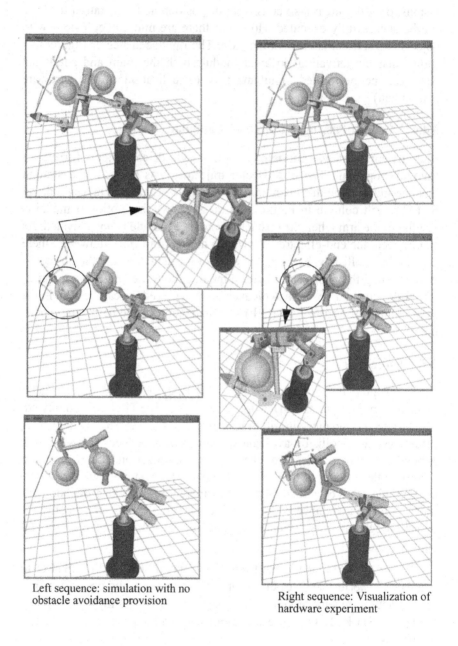

Left sequence: simulation with no
obstacle avoidance provision

Right sequence: Visualization of
hardware experiment

Figure 3.21 Collision avoidance with stationary spherical objects

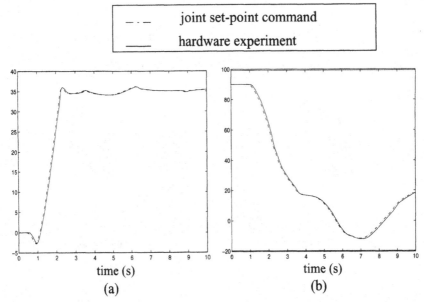

Figure 3.22 Case 2: a) joint 2, b) joint 3 (degrees)

- **Position and orientation tracking**: Considering the complexity of the singular regions existing in the 3D workspace of a 7-DOF manipulator, the singularity-robustness formulation of redundancy was shown to be necessary in practical applications. It was shown that by a proper selection (or a time-varying formulation) of W_v, the weighting matrix of the singularity-robustness task, the effect of this term on tracking performance can be minimized.

- **Performing additional task(s)**: Joint limit avoidance and obstacle avoidance were implemented for REDIESTRO. It was shown that the formulation of additional tasks as inequality constraints, may result in rapid change in joint velocities causing a large pulse in joint accelerations. In a practical implementation, since the maximum acceleration of each joint would be limited, such a commanded joint acceleration would result in saturation of the actuators. A time-varying formulation of the weighting matrix, W_c, was proposed which successfully overcame this problem.

- Fine tuning of control gains and weighting matrices

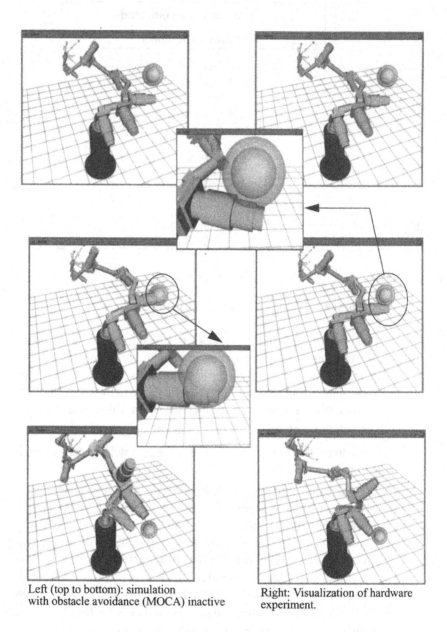

Left (top to bottom): simulation
with obstacle avoidance (MOCA) inactive

Right: Visualization of hardware
experiment.

Figure 3.23 Collision Avoidance with moving spherical object.

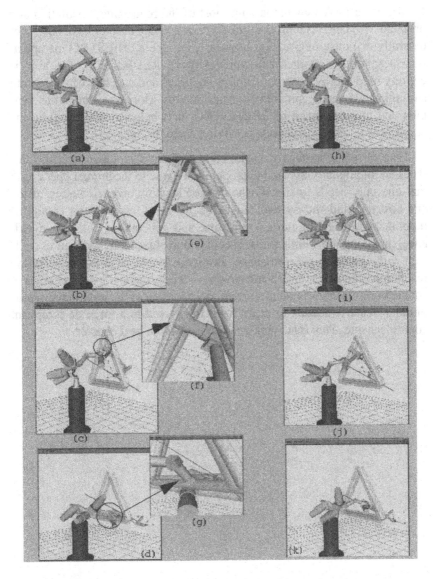

Left sequence: Simulation with obstacle avoidance inactive.

Right sequence: Visualization of the hardware demonstration with obstacle avoidance active

Figure 3.24 Passing through a triangular opening

Three scenarios encompassing most of the redundancy-resolution and obstacle-avoidance system features described in this chapter have been successfully demonstrated on real hardware, i.e., the REDIESTRO manipulator. Despite the geometrical complexity of REDIESTRO, the arm is entirely modeled by decomposition of the links and attached actuators into sub-links modeled by simple volume primitives. Moreover, due to the complex and unusual shape of REDIESTRO, it is believed that adapting the algorithms to other manipulators will in general be simpler.

The current redundancy-resolution and obstacle-avoidance scheme provides an intelligently assisted tele-operation mode to the human operator in that one only needs to specify the desired location and orientation of the end-effector, and the system automatically takes care of the details of motion control, configuration selection, and generalized collision avoidance, including joint-limit and self-collision avoidance, in addition to collision with objects in the workspace. However, at this stage the redundancy-resolution scheme cannot handle situations where the manipulator comes in contact with its environment. Further modification to the redundancy-resolution scheme is needed in order for it to be used in a force or compliant control scheme. This issue will be addressed in the next chapter.

4 Contact Force and Compliant Motion Control

4.1 Introduction

Robotic tasks mainly fall into two categories: Constrained and unconstrained motions. During the initial stages of development in robotics, most successful applications dealt with position control of unconstrained motion of robot manipulators. The nature of these tasks does not require a robot to come in contact with its environment (work piece). Spray painting is an example of such a task in which the robot brings a spray gun near the surface to be painted and then sweeps across the surface with a specified velocity. Another example is that of seam welding. In some applications, where a robot comes in contact with its environment (as in the case of material handling), precise control of the interaction with the object is not required. The problem that arises when using a position control scheme in a constrained motion is that the robot-environment interaction forces are treated as disturbances. The controller tries to reject these forces, and hence, gives rise to larger interaction forces. The consequences of this are saturation, instability, or even physical failure and damage to the robot and the environment. Whitney [94] gives a historical perspective on robot force control. Force control strategies have been mainly designed to use force feedback sensory information.

Salisbury [60] proposed a stiffness control scheme. Raibert and Craig [56] proposed a hybrid position-force control scheme. Yoshikawa [96], McClamroch and Wang [45] proposed a method based on a constrained dynamic model of a manipulator. Hogan introduced the impedance control idea in a series of papers in the mid-1980's. In [30], he proposed the fundamental theory of impedance control which showed that command and control of a vector such as position or force is not enough to control the dynamic interactions between a manipulator and its environment. This emphasizes the main problem of hybrid position-force control, i.e., its failure to recognize the importance of manipulator impedance. The impedance control scheme overcomes this problem, but it ignores the distinction between position and force controlled subspaces, and no attempt is made to follow a commanded force trajectory. Therefore, Anderson and Spong [1] proposed a *Hybrid Impedance Control* (HIC) scheme, and Liu and Goldenberg [40] introduced a robust HIC method.

The aforementioned methods can be divided into two main categories, referred to as constrained motion [56], [96], [45], and compliant motion [30], [1], [40] approaches. In the next sections, an outline of these approaches is given. Note that the above mentioned algorithms are not directly applicable to redundant manipulators. However, a careful review of these algorithms gives guidelines for selecting force or compliant motion control for redundant manipulators. Recent work has specifically concentrated on force o compliant motion control for redundant manipulators [69], [53], [50], [29]. A class of nonlinear contact controllers is introduced in [69]. Each controller consists of a nonlinear gain cascaded with a linear fixed-gain proportional-integral (PI) force controller and proportional-derivative (PD) compliance controller. In [53], an extended HIC scheme is presented which achieves an inertial decoupling of the motion and force controlled subspaces and internal motion control using a minimal parametrization of motion and force controlled subspaces and the null-motion component. No experimental results are given. A force control scheme for redundant manipulators is presented in [50] which decouples the motion of the manipulator into task-space motion and internal motion while providing for the selection of the dynamic characteristics for the motions. Hattori and Ohnishi [29] describe a decentralized compliant motion control scheme for redundant manipulators based on the concept of virtual impedance. The manipulator is divided into several subsystems each of which performs autonomously using virtual impedance and information from the end-effector subsystem. Simulation and experimental results are given for a redundant planar manipulator.

In the remainder of this chapter, algorithms proposed for force and compliant motion control of redundant manipulators are presented. Section 4.3.1 addresses the extension of configuration control at the acceleration level. Section 4.3.2 introduces the Augmented Hybrid Impedance Control (AHIC) scheme. The feasibility of this scheme with respect to performing both the main and additional tasks is studied using a 3-DOF planar arm. The AHIC scheme is then modified to cop with the uncontrolled self-motion. The AHIC scheme with self-motion stabilization is presented in 4.3.3. An adaptive version of the AHIC scheme is presented in Section

4.2 Literature Review

4.2.1 Constrained Motion Approach

This approach considers the control of a manipulator constrained by a rigid object[1] in its environment. If the environment imposes purely kinematic constraints on the end-effector motion, only a static balance of forces and torques occurs (assuming that the frictional effects are neglected). This implies no energy transfer or dissipation between the manipulator and the environment. This underlies the main modeling assumption made by [45] where an algebraic vector equation restricts the feasible end-effector poses. The constrained dynamics can be written as:

$$H(q)\ddot{q} + h(q, \dot{q}) = \tau - J^T F$$

$$\Phi(p) = 0$$

(4.2.1)

where τ is the vector of applied forces (torques), $H(q)$ is the $n \times n$ symmetric positive definite inertia matrix, h is the vector of centrifugal, Coriolis, and gravitational torques. $p \in R^n$ is the generalized task coordinates, and $\Phi(p) \in R^m$ is the constraint equation, continuously differentiable with respect to p. It is assumed that the Jacobian matrix is square and of full rank. The analysis given below follows that in [45], the generalized force[2] F in (4.2.1) is given by:

$$F = \left(\frac{\partial \Phi(p)}{\partial p}\right)^T \lambda$$

(4.2.2)

where $\lambda \in R^{m \times 1}$ is the vector of generalized Lagrange multipliers. Using the forward kinematic relations:

$$\dot{p} = J\dot{q}$$

$$\ddot{p} = J\ddot{q} + \dot{J}\dot{q}$$

(4.2.3)

---

1. *A work environment or object is said to be rigid when it does not deform as a result of application of generalized forces by the manipulator.*
2. *In the rest of this chapter, the term "force" refers to both interaction force and torque.*

and assuming that the Jacobian matrix is invertible, we can obtain the following constrained dynamics expressed with respect to generalized task coordinates directly from (4.2.1):

$$H_p(p)\ddot{p} + h_p(p, \dot{p}) = u - F$$

$$\Phi(p) = 0 \tag{4.2.4}$$

where

$$H_p = J^{-T}H(q)J^{-1}$$

$$h_p = -H_p\dot{J}\dot{q} + J^{-T}h(q, \dot{q}) \tag{4.2.5}$$

$$u = J^{-T}\tau$$

A nonlinear transformation can then be used to transfer to a new coordinate frame. It is assumed that there is an open set $\Theta \subset R^{n-m}$ and a function Ω such that

$$\Phi(\Omega(p_2), p_2) = 0 \qquad \forall p_2 \in \Theta \tag{4.2.6}$$

where

$$p = \begin{bmatrix} p_{1_{m \times 1}} \\ p_{2_{(n-m) \times 1}} \end{bmatrix} \tag{4.2.7}$$

Now, defining another coordinate system represented by the vector x, we obtain the following nonlinear transformation X:

$$x = X(p) = \begin{bmatrix} p_1 - \Omega(p_2) \\ p_2 \end{bmatrix}$$

which is differentiable and has a differentiable inverse given by:

$$p = Q(x) = \begin{bmatrix} x_1 + \Omega(x_2) \\ x_2 \end{bmatrix} \tag{4.2.8}$$

where x is partitioned conformably with (4.2.7). The Jacobian of (4.2.8) is defined by:

$$T(x) = \frac{\partial Q(x)}{\partial x} = \begin{bmatrix} I_m & \dfrac{\partial \Omega(x_2)}{x_2} \\ 0 & I_{n-m} \end{bmatrix} \quad (4.2.9)$$

Transforming the equation of motion in (4.2.4) to the generalized coordinate x, we obtain:

$$H_x(x)\ddot{x} + h_x(x, \dot{x}) = T^T u - T^T F$$
$$x_1 = 0 \quad (4.2.10)$$

where

$$H_x = T^T(x) H_p(Q(x)) T(x) \quad (4.2.11)$$

$$h_x = T^T(x) H_p(Q(x)) \dot{T}(x) \dot{x} + T^T(x) h_p(Q(x), T(x)\dot{x})$$

Note that in this transformed frame, the constraint equation takes the simple form $x_1 = 0$. Equations (4.2.10) can be simplified as follows:

$$E_1 H_x E_2^T \ddot{x}_2 + E_1 h_x = E_1 T^T(u - F)$$
$$E_2 H_x E_2^T \ddot{x}_2 + E_2 h_x = E_2 T^T u \quad (4.2.12)$$
$$x_1 = 0$$

where E_1 and E_2 are defined by

$$I_n = [E^T{}_1, E^T{}_2]$$
$$E^T{}_1 = \begin{pmatrix} I_m \\ 0 \end{pmatrix} \qquad E^T{}_2 = \begin{pmatrix} 0 \\ I_{n-m} \end{pmatrix} \quad (4.2.13)$$

The hybrid control law is defined as

$$T^T u = u_x + u_f \tag{4.2.14}$$

where

$$u_x = H_x \begin{bmatrix} 0 & E_2^T \end{bmatrix} [\ddot{x}_d + K_v(\dot{x}_d - \dot{x}) + K_p(x_d - x)] + h_x(x, \dot{x})$$

$$u_f = \begin{bmatrix} E_1^T & 0 \end{bmatrix} [T^T F_d + G_F T^T (F_d - F)]$$

$$\tag{4.2.15}$$

where K_p, K_v, and G_F are feedback gain matrices. By replacing the control law (4.2.14) in the equations of motion (4.2.12), the following closed-form system of equations is obtained:

$$E_1 H_x E_2^T(\ddot{e}_2 + K_v \dot{e}_2 + K_p e_2) = (I_m + G_F)E_1 T^T(F_d - F)$$

$$E_1 H_x E_2^T(\ddot{e}_2 + K_v \dot{e}_2 + K_p e_2) = 0$$

$$e_1 = 0$$

$$\tag{4.2.16}$$

where $e_1 = x_1 - x_{1d}$ and $e_2 = x_2 - x_{2d}$. The closed-loop equations of motion given by (4.2.16) imply that $e_2 \to 0$ as $t \to \infty$ through a proper choice of feedback gains and also $F \to F_d$ as $t \to \infty$. Hence, the closed-loop system is asymptotically stable.

A hybrid position and force controller is proposed in [56] where the task space is divided into two orthogonal subspaces - position controlled and force-controlled - using a selection matrix S. The diagonal elements of the selection matrix S are selected as 0 or 1 depending on which degrees of freedom are force-controlled and which are position-controlled (Figure 4.1).

Mills [46] showed that the constrained motion control approach of McClamroch and Wang [45] is identical to the hybrid position and force control scheme if the selection matrix S is replaced by:

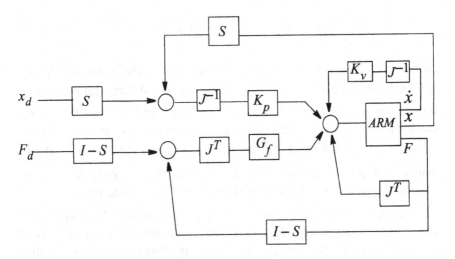

Figure 4.1 Schematic diagram of the hybrid position and force controlled system

$$S = [0 \quad E_2^T]$$

$$I - S = [E_1^T \quad 0]$$

(4.2.17)

Note that these methods are not directly applicable to redundant manipulator.

4.2.2 Compliant Motion Control

In contrast to the constrained motion approach, compliant motion control as its name implies, deals with a compliant environment. This approach is aimed at developing a relationship between interaction forces and a manipulator's position instead of controlling position and force independently. This approach is limited by the assumption of small deformations of the environment, with no relative motion allowed in coupling. Salisbury [60] proposed the stiffness control method. The objective is to provide a stabilizing dynamic compensator for the system such that the relationship between the position of the closed-loop system and the interaction forces is constant over a given operating frequency range. This can be written mathematically as follows:

$$\delta F(j\omega) = K\delta X(j\omega), \qquad 0 < \omega < \omega^o \qquad (4.2.18)$$

where $\delta F(j\omega)$ is the $n \times 1$ vector of deviations of the interaction forces and torques from their equilibrium values in a global Cartesian coordinate frame; $\delta X(j\omega)$ is the $n \times 1$ vector of deviations of the positions and orientations of the end-effector from their equilibrium values in a global Cartesian coordinate frame; K is the $n \times n$ real-valued nonsingular stiffness matrix; and ω^o is the bandwidth of operation. By specifying K, the user governs the behavior of the system during constrained maneuvers.

Hogan [30] proposed the impedance control idea. Impedance control is closely related to stiffness control. However, stiffness is merely the static component of a robot's output impedance. Impedance control goes further and attempts to modulate the dynamics of the robot's interactive behavior. The main idea of impedance control is to make the manipulator act as a mass-spring-dashpot system in each degree of freedom in its workspace. Therefore, the manipulator is seen as an apparent impedance given by:

$$M^d(\ddot{X} - \ddot{X}^d) + B^d(\dot{X} - \dot{X}^d) + K^d(X - X^d) = -F^e \qquad (4.2.19)$$

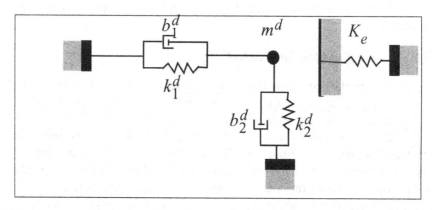

Figure 4.2 Apparent impedance of a manipulator in each degree of freedom in task space

where M^d, B^d, and K^d are diagonal $m \times m$ matrices of the desired mass, damping, and stiffness; F^e is the vector of the environmental reaction forces; and the superscript d refers to desired values.

First, let us define the operational-space dynamic equation of motion of the manipulator[1] as:

$$H_x(X)\ddot{X} + h_x(X, \dot{X}) = J^{-T}u + F^e \tag{4.2.20}$$

where H_x is the Cartesian inertia matrix, and h_x is the vector of centrifugal, Coriolis, and gravity terms acting in operational space. Then as proposed in [1], an inner and outer loop control strategy (Figure 4.3) can be used to achieve the closed-loop dynamics specified by (4.2.19)

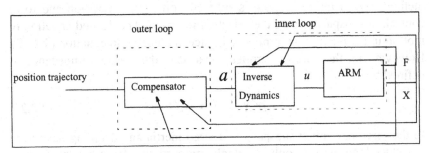

Figure 4.3 Inner-outer loop control strategy

In the absence of uncertainties in the dynamic parameters of the manipulator, the inner loop is a feedback linearization loop of the form

$$u = J^T(H_x a + h_x - F^e) \tag{4.2.21}$$

which results in the double integrator system $\ddot{X} = a$. The output of the outer loop is a target acceleration obtained by solving (4.2.19):

$$a = \ddot{X}^d - M^{d^{-1}}[B^d(\dot{X} - \dot{X}^d) + K^d(X - X^d) - F^e] \tag{4.2.22}$$

---

1. *If we consider a non-redundant manipulator not in a singular configuration, then*

$$H_x = J^{-T}H_q J^{-1}, \qquad h_x = J^{-T}h_q - H_x \dot{J}\dot{q}$$

Hogan indicated that the impedance control scheme is capable of controlling the manipulator in both free space and constrained maneuvers while eliminating the switching between free-motion and constrained motion controllers.

A typical compliant motion task is the surface cleaning scenario shown in Figure 4.4. As we can see a target trajectory is defined to be identical to the desired trajectory in free motion. However, in order to maintain contact with the environment, the target trajectory is defined to be different from the desired trajectory in constrained maneuvers. Depending on the desired impedance characteristics and the environment, the robot will follow an actual path which results in a certain contact force with the environment.

It should be noted that in the impedance control scheme, no attempt is made to follow a commanded force trajectory. To overcome this problem, Anderson and Spong [1] proposed a Hybrid Impedance Control (HIC) method. Again the task space is split into orthogonal position and force controlled subspaces using the selection matrix S. The desired equation of motion in the position-controlled subspace is identical to equation (4.2.19). However, in the force-controlled subspace, the desired impedance is defined by:

$$M^d \ddot{X} + B^d \dot{X} - F^d = -F^e \tag{4.2.23}$$

In the force-controlled subspace, a desired inertia and damping have been introduced because if only a simple proportional force feedback were applied, the response could be very under-damped for an environment with high stiffness. In the case of loss of contact with the environment or approaching the surface ($F^e = 0$), equation (4.2.23) becomes

$$M^d \ddot{X} + B^d \dot{X} = F^d \tag{4.2.24}$$

If we assume a constant desired force, positive diagonal inertia and damping matrices, and $\dot{X}(0) = 0$, then the ith component of the velocity vector $\dot{X}$ is given by:

$$\dot{X}_i(t) = \frac{F_i^d}{B_i^d}(1 - e^{-(B_i^d/M_i^d)t}) \tag{4.2.25}$$

Therefore

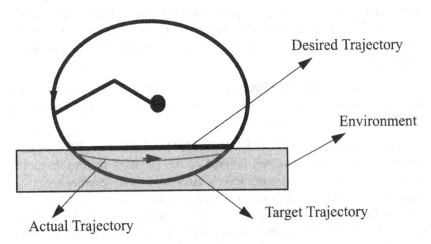

Desired Trajectory

Environment

Target Trajectory

Actual Trajectory

Figure 4.4 Surface cleaning using impedance controller

$$\left|\dot{X}_i(t)\right| < \frac{F_i^d}{B_i^d} \qquad and \qquad \lim_{t \to \infty} \dot{X}_i(t) = \frac{F_i^d}{B_i^d} \qquad (4.2.26)$$

This guarantees that the arm approaches the environment with a velocity that can be properly limited in order to reduce impact forces.

Again, note that these methods are not directly applicable to redundant manipulators. The main reasons are the use of the Cartesian model of manipulator dynamics, and calculation of the command input in task space. As we mentioned earlier, for a redundant manipulator, the task space requirements cannot uniquely determine joint space configurations. An augmented hybrid impedance controller which overcomes this problem will be proposed in next section.

4.3 Schemes for Compliant and Force Control of Redundant Manipulators

The problem of compliant motion control of redundant manipulators has not attained the maturity level of its non-redundant counterpart. There is little work that addresses the problem of redundancy resolution in a compliant motion control scheme. There are two major issues to be addressed in extending existing compliant motion schemes to the case of redundant manipulators:

(i) The nature of compliant motion control requires expressing the manipulator's task in Cartesian space; therefore, such schemes are usually based on the Cartesian dynamic model of manipulator. However, in the presence of redundancy, there is not a unique map from Cartesian space to joint space.

(ii) Most redundancy resolution techniques are at the velocity level, and simple extensions of these techniques to the acceleration level have resulted in the self-motion phenomenon.

For instance, Gertz et al. [23], Walker [91] and Lin et al. [39] have used a generalized inertia-weighted inverse of the Jacobian to resolve redundancy in order to reduce impact forces. However, these schemes are single purpose algorithms, and cannot be used to satisfy additional criteria. An extended impedance control method is discussed in [2] and [51]; the former also includes an HIC scheme. These schemes can be considered as multipurpose algorithms since different additional tasks can be incorporated in HIC without modifying the schemes and the control laws. However, there are two major drawbacks to these schemes: (i) The dimension of the additional task should be equal to the degree of redundancy, which makes the approach not applicable for a wide class of additional tasks, i.e., additional tasks that are not active for all time such as obstacle avoidance in a cluttered environment. (ii) The HIC scheme introduces the possibility of controlling tasks either by a position controlled or a force controlled scheme. The possibility of having an additional task controlled by a force controlled scheme is ignored by including the additional task in the position controlled subspace of the extended task. Shadpey et al. [72] have proposed an Augmented Hybrid Impedance Control (AHIC) scheme to overcome these problems (see Section 4.3.2). This scheme enjoys the following major advantages:

(i) Different additional tasks can be easily incorporated in the AHIC scheme without modifying the scheme and the control law.

(ii) An additional task can be included in the force-controlled subspace of the augmented task. Therefore, it is possible to have a multiple-point force control scheme.

(iii) Task priority and singularity robustness formulation of the AHIC scheme relaxes the restrictive assumption of having a non-singular augmented Jacobian.

However, the scheme in [72] exhibits the self-motion phenomenon, i.e., motion of the arm in the null space of the Jacobian. Another AHIC scheme which has the above mentioned characteristics [73] is presented in Section 4.3.3. Moreover, by modifying both the inner and outer control loops, the self-motion is damped when the dimension of the augmented task is smaller than that of the joint space. This scheme is also more amenable to an adaptive implementation. An adaptive version of the AHIC scheme [74] is described in Section 4.3.4 .

4.3.1 Configuration Control at the Acceleration Level

Similar to the pseudo-inverse solution given by (2.3.30), the following weighted damped least-squares solution can be obtained:

$$
\ddot{q} = [J_e^T W_e J_e + J_c^T W_c J_c + W_v]^{-1} [J_e^T W_e (\ddot{X} - \dot{J}_e \dot{q})
$$
$$
+ J_c^T W_c (\ddot{Z} - \dot{J}_c \dot{q})]
$$

(4.3.1)

This minimizes the following cost function:

$$
L = \ddot{E}_e^T W_e \ddot{E}_e + \ddot{E}_c^T W_c \ddot{E}_c + \ddot{q}^T W_v \ddot{q}
$$

(4.3.2)

where

$$
\ddot{E}_e = \ddot{X}^d - (\ddot{X} - \dot{J}_e \dot{q}) \qquad and \qquad \ddot{E}_c = \ddot{Z}^d - (\ddot{Z} - \dot{J}_c \dot{q}) \quad (4.3.3)
$$

However, this solution is incapable of controlling the null space component of joint velocities (see Section 2.3.2). A remedy for this difficulty is to differentiate the configuration control solution at the velocity level given by equation (2.3.19). This yields

$$
\ddot{q} = [J_e^T W_e J_e + J_c^T W_c J_c + W_v]^{-1} [A + B]
$$

(4.3.4)

where

$$
A = J_e^T W_e (\ddot{X} - \dot{J}_e \dot{q}) + J_c^T W_c (\ddot{Z} - \dot{J}_c \dot{q})
$$

$$
B = \dot{J}_e^T W_e (\dot{X} - J_e \dot{q}) + \dot{J}_c^T W_c (\dot{Z} - J_c \dot{q})
$$

Therefore, following the reference joint velocity given by equation (2.3.19) and the acceleration trajectory given by (4.3.4), we get a special solution that minimizes the joint velocities when $k < r$, i.e., there are not as many active tasks as the degree of redundancy, and we have the best solution in the least-squares sense when $k > r$. In all cases the presence of W_v ensures the boundedness of the joint velocities.

4.3.2 Augmented Hybrid Impedance Control using the Computed Torque Algorithm

The *AHIC* scheme, shown in Figure 4.5, can be broken down into two different control loops. The outer loop generates an *Augmented Cartesian Target Acceleration (ACTA)* trajectory reflecting the desired impedance in the position-controlled subspaces, and the desired force in the force-controlled subspaces of the main and additional tasks. From this point of view, the AHIC problem can be formulated as that of tracking an ACTA trajectory which is generated in real time. The inner-loop control problem consists of selecting an input τ to the actuators which makes the end-effector track a desired trajectory generated by the outer loop.

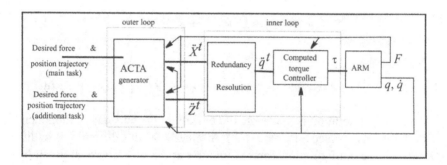

Figure 4.5 Block diagram of the AHIC scheme using the computed torque controller

4.3.2.1   Outer-loop design

The design of the outer-loop part of the AHIC scheme is described in this section. As mentioned in Section 4.2, the idea is to split the spaces corresponding to the main task X and additional task Z into position- and force-controlled subspaces. Impedance control is used in the position-controlled subspace. Therefore, the objective is to make the manipulator act as a mass-spring-dashpot system with desired inertia, damping, and stiffness in each dimension of the position-controlled subspace of the main and additional tasks. In the force-controlled subspace, a desired inertia and damping have been introduced because, if only a simple proportional force feedback were applied, the response could be very under-damped for an environment with high stiffness.

The motion of the manipulator in both subspaces can be expressed by a single matrix equation using selection matrices S_x and S_z, as follows:

$$M_x^d(\ddot{X} - S_x\ddot{X}^d) + B_x^d(\dot{X} - S_x\dot{X}^d) + K_x^d S_x(X - X^d)$$
$$-(I - S_x)F_x^d = -F_x^e \quad (a)$$

$$M_z^d(\ddot{Z} - S_z\ddot{Z}^d) + B_z^d(\dot{Z} - S_z\dot{Z}^d) + K_z^d S_z(Z - Z^d)$$
$$-(I - S_z)F_z^d = -F_z^e \quad (b)$$

$$(4.3.5)$$

where the superscript d denotes the desired values; the subscripts x and z refer to the main and additional tasks respectively; the diagonal matrices M, B, and K are the desired mass, damping, and stiffness matrices; F^d and $-F^e$ are vectors of the desired and the environmental reaction forces; and S_x and S_z are the diagonal selection matrices which have 1's on the diagonal for position-controlled subspaces and 0's for the force-controlled subspaces.

Solving the motion equations (4.3.5) for the accelerations $\ddot{X}$ and $\ddot{Z}$ leads to the *Cartesian Target Acceleration (CTA)* trajectories of the main, $\ddot{X}^t$, and additional tasks, $\ddot{Z}^t$:

$$\ddot{X}^t = S_x\ddot{X}^d - (M_x^d)^{-1}[B_x^d(\dot{X} - S_x\dot{X}^d) + K_x^d S_x(X - X^d)$$
$$-(I - S_x)F_x^d + F_x] \quad (a)$$

$$(4.3.6)$$

$$\ddot{Z}^t = S_z\ddot{Z}^d - (M_z^d)^{-1}[B_z^d(\dot{Z} - S_z\dot{Z}^d) + K_z^d S_z(Z - Z^d)$$
$$-(I - S_z)F_z^d + F_z^e] \quad (b)$$

Now the AHIC scheme can be formulated to track the *CTA* trajectories. Using the configuration control approach - equation (4.3.1), the desired *Joint Target Acceleration (JTA)* trajectory ($\ddot{q}^t$) can be found by replacing the *CTA* trajectories of the main and additional tasks in Equation (4.3.1).

$$\ddot{q}^t = [J_e^T W_e J_e + J_c^T W_c J_c + W_v]^{-1}[J_e^T W_e(\ddot{X}^t - \dot{J}_e\dot{q})$$
$$+ J_c^T W_c(\ddot{Z}^t - \dot{J}_c\dot{q})]$$

$$(4.3.7)$$

Remark: Duffy [20] has indicated that in the case of compliant motion of a manipulator in 3D space, the end-effector velocities (linear, angular) and forces (forces, torques) should be considered as screws represented in axis and ray coordinates. Therefore, in general the concept of orthogonality of force and position controlled subspaces is not valid. As shown in [80], the concept that is appropriate is that of "reciprocal" subspaces, i.e., the set of motion screws should be decomposed into mutually reciprocal free and constraint subspaces.

4.3.2.2   Inner-loop

The dynamics of a rigid manipulator in the absence of disturbances are described by:

$$H(q)\ddot{q} + h(q, \dot{q}) + G(q) + f(\dot{q}) = \tau + J_e^T F_x^e + J_{c1}^T F_z^e \qquad (4.3.8)$$

where τ is the vector of applied forces (torques), $H(q)$ is the $n \times n$ symmetric positive definite inertia matrix, h is the vector of centrifugal and Coriolis forces, f is the vector of frictional forces, and G is the vector of gravitational forces. The last term on the right-hand side of the equation is only needed if another point of the manipulator (other than the end-effector) is in contact with the environment; F_z^e denotes the reaction force corresponding to a second constraint surface, and J_{c1} is the Jacobian of the contact point.

As mentioned earlier, the responsibility of the inner loop is to ensure that the manipulator tracks the JTA trajectory. Referring to the dynamic equation of the manipulator, the input torque is selected by:

$$\tau = H\ddot{q}^t + h(q, \dot{q}) + G(q) + f(\dot{q}) - J_e^T F_x^e - J_{c1}^T F_z^e \qquad (4.3.9)$$

which guarantees perfect following of the JTA trajectory in the absence of uncertainties in the manipulator's parameters.

4.3.2.3   Simulation Results for a 3-DOF Planar Arm

The performance of the AHIC scheme has been studied using simulations involving a 3-revolute-joint planar manipulator (Figure 4.6). In all cases, a constraint surface is defined by the position P_c and orientation θ_c of a frame C attached to this surface. The main task (same for all cases), defined in the constraint frame, is specified by a desired impedance (inertia, damping, and stiffness) in tracking of the desired position trajectory in the X_c direction, and desired force trajectory in the Y_c direction. The selected values for the simulations are: $m^d=1$, $b^d=120$, and $k^d=3600$. The environment is modelled as a spring with stiffness $K_e=10000$ N/m. The desired position trajectory is calculated by linear interpolation between the initial and final points (constant velocity trajectory), and the force trajectory is defined by $f^d = -100$ N. For each individual case, a different additional task is defined. A general block diagram of the simulation is shown in Figure 4.7 where T denotes the homogenous coordinate transformation between two different frames (W refers to the workspace, and C refers to the end-effector constraint surface). Note that the blocks shown by dashed lines are needed if the only the additional task is force-controlled (CI refers to the second constraint surface).

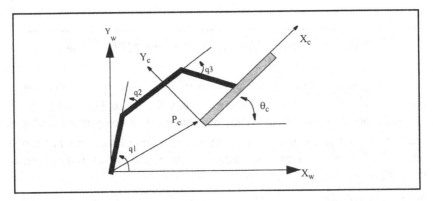

Figure 4.6 3-DOF planar manipulator used in the simulation

Joint Limit Avoidance (JLA)

The formulation of the additional task was given in Section 2.4.1 . In the first simulation, JLA is inactive, and the resulting errors in the position and force controlled subspaces () both converge to small values (practically zero). However, the joint three value goes below -100 degrees. In the second simulation, JLA is active and the minimum limit for joint three is selected as -80 degrees. The simulation results again show that the force and position trajectories are tracked correctly, and also the limit on joint three is respected.

Static and Moving Obstacle Avoidance (SOCA and MOCA)

The formulation of the additional task was given in Section 2.4.2 . The results for SOCA are indicated in , When the obstacle avoidance algorithm is inactive, the main task trajectories are followed correctly. However, a collision occurs. By activating obstacle avoidance, the collision is avoided and the main task requirement is also satisfied.

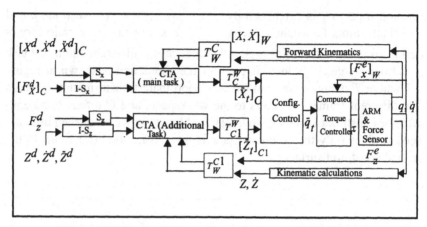

Figure 4.7 General block diagram of the AHIC scheme

In the next simulation, the position of the tip in the X_c direction is required to be fixed, while exerting a constant force equal to -100 N in the Y_c direction. shows that the main task has been accomplished within a short time, and from this time onwards, the manipulator does not move until the MOCA additional task becomes active, and successfully prevents the collision.

Task Compatibility

The objective of this additional task is to position the arm in the posture which requires minimum torque for a desired force in a certain direction. The formulation of this additional task is given in Section 2.4.3 .

Figure 4.11 shows the results of the simulation for this case. The main task consists of keeping the manipulator tip at a fixed position in the X direction while exerting -100 N in the Y direction. As we can see in Figure 4.11b, the manipulator reconfigures itself to find the posture which requires the minimum torque to exert the desired force. Figure 4.11c shows how the value of the objective function - task compatibility index given by (2.4.16) - increases to reach the optimal configuration. Figure 4.11d shows the force ellipsoid for the initial and final configurations. Note that the force transfer ratio along the Y direction has been increased. Figures 4.11e and f show that the force and position trajectories of the main task were followed correctly. Note that the required torque is reduced when the additional task is active (Figure 4.11g).

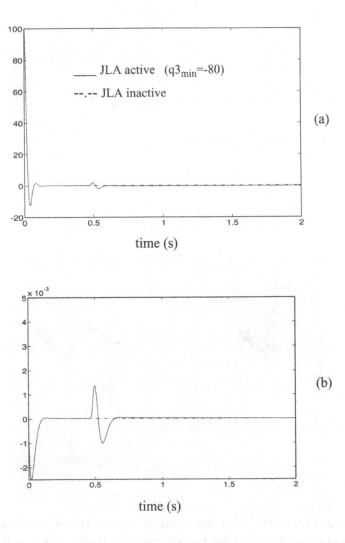

Figure 4.8 Simulation results for the AHIC scheme with Joint Limit Avoidance: (a) force error (N); (b) position error (mm)

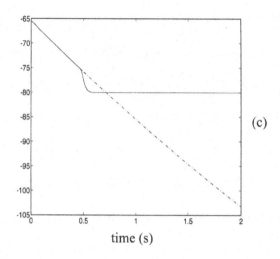

(c)

time (s)

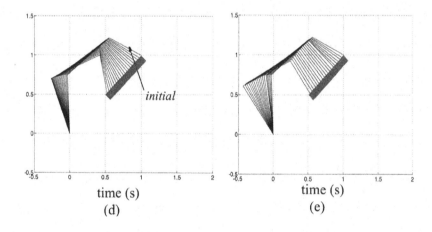

time (s)   time (s)
(d)   (e)

Figure 4.8 (contd.) Simulation results for the AHIC scheme with Joint Limit Avoidance: (c) joint 3 variable (deg); (d) robot motion - JLA inactive; (e) robot motion - JLA active

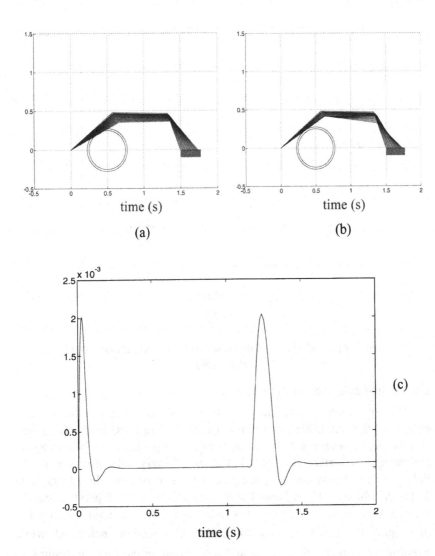

time (s)

(a)

time (s)

(b)

(c)

time (s)

Figure 4.9 Static Obstacle Collision Avoidance: (a) robot motion - SOCA off; (b) robot motion - SOCA on; (c) position error (m)

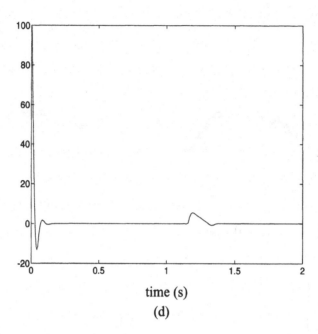

time (s)

(d)

Figure 4.9 (contd.) Static Obstacle Collision Avoidance: (d) force
error (N)

Force-Controlled Additional Task

We have already noted that the additional task(s) can be included in
either position-controlled or force-controlled subspaces. In the following
simulation, the additional task consists of exerting a constant force to a second compliant surface (Figure 4.12) by an arbitrary point Z on one of the
links - in this simulation, the joint between the second and third links, joint
3. The Jacobian of the additional task is the Jacobian of the point Z, and the
desired force in the Y_{c1} direction is to be specified. The main task consists
of keeping the position of the tip in the X_w direction unchanged, while
exerting a constant -100 N force in Y_W direction on the first constraint surface. The additional task is to exert a 100 N force (in the Y_{c1} direction) on
the second constraint surface by joint three. Figure 4.13b shows the motion
of the joints and Figures 4.13c, and d show that the main task is executed
correctly. Figure 4.13e shows that the desired force is exerted on the second
constraint surface. Note that, although initially joint three is not in contact
with the second constraint surface, the AHIC scheme works correctly and
makes this point move toward the surface with a bounded velocity.

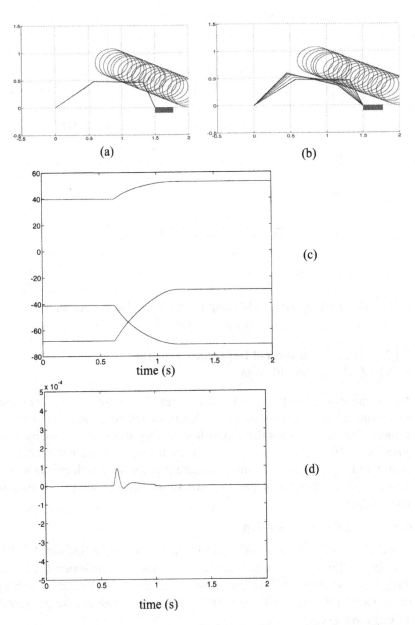

Figure 4.10 Moving Obstacle Collision Avoidance: (a) robot motion -
MOCA off; (b) robot motion - MOCA on; (c) joint variables (deg);
(d) position error (m).

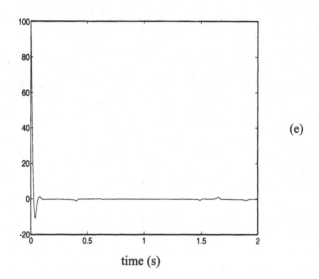

(e)

Figure 4.10 (contd.) Moving Obstacle Collision Avoidance:
(e) force error (N)

4.3.3   Augmented Hybrid Impedance Control
with Self-Motion Stabilization

As we mentioned earlier, redundancy resolution at the acceleration
level is aimed at minimizing joint accelerations and not controlling the self-
motion of the arm. This is the major shortcoming of the AHIC scheme proposed in Section 4.3.2. In this section by modifying both the inner and outer
control loops, a new AHIC control scheme is proposed which enjoys all the
desirable characteristics of the previous scheme and achieves self-motion
stabilization.

4.3.3.1   Outer-Loop Design

The design of the outer-loop is similar to the design in Section 4.3.2.1.
The only difference is that instead of calculating an *Augmented Cartesian
Target Acceleration (ACTA)* trajectory, we describe the desired motion by
an *Augmented Cartesian Target (ACT)* trajectory at position, velocity, and
acceleration levels.

The motion of the manipulator in both subspaces can be expressed by a
single matrix equation using the selection matrices S_x and S_z, as follows:

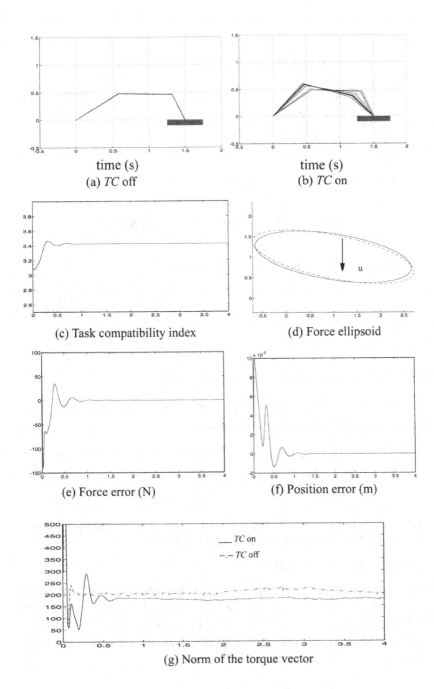

(a) *TC* off

(b) *TC* on

(c) Task compatibility index

(d) Force ellipsoid

(e) Force error (N)

(f) Position error (m)

(g) Norm of the torque vector

Figure 4.11 Task compatibility simulation results

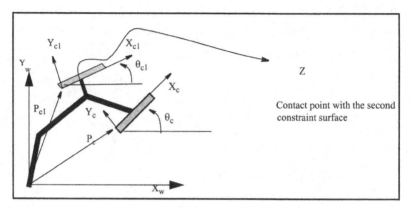

Figure 4.12 Force-controlled additional task

$$M_x^d(\ddot{X}^t - S_x\ddot{X}^d) + B_x^d(\dot{X}^t - S_x\dot{X}^d) + K_x^d S_x(X^t - X^d)$$
$$- (I - S_x)F_x^d = -F_x^e \quad \text{(a)}$$
$$\quad\quad\quad\quad\quad\quad\quad\quad\quad (4.3.10)$$
$$M_z^d(\ddot{Z}^t - S_z\ddot{Z}^d) + B_z^d(\dot{Z}^t - S_z\dot{Z}^d) + K_z^d S_z(Z^t - Z^d)$$
$$- (I - S_z)F_Z^d = -F_Z^e \quad \text{(b)}$$

where the same definitions as in (4.3.5) are used.

The ACT trajectory $[X^{t^T}, Z^{t^T}]^T$ is the unique solution of the differential equations (4.3.10) with initial conditions:

$$X^t(0) = X^d(0) \quad\quad \dot{X}^t(0) = \dot{X}^d(0)$$
$$\quad\quad\quad\quad\quad\quad\quad\quad (4.3.11)$$
$$Z^t(0) = Z^d(0) \quad\quad \dot{Z}^t(0) = \dot{Z}^d(0)$$

Notice that the presence of measurement forces in these equations requires that the ACT trajectory should be generated online.

4.3.3.2 Inner-Loop Design

The dynamics of a rigid manipulator are described by equation (4.3.8). The controller should be designed to calculate the torque input to the dynamic equation (4.3.8), which ensures the tracking of the ACT trajectory. The procedure is as follows: First, a Cartesian reference trajectory is defined for both the main and additional tasks:

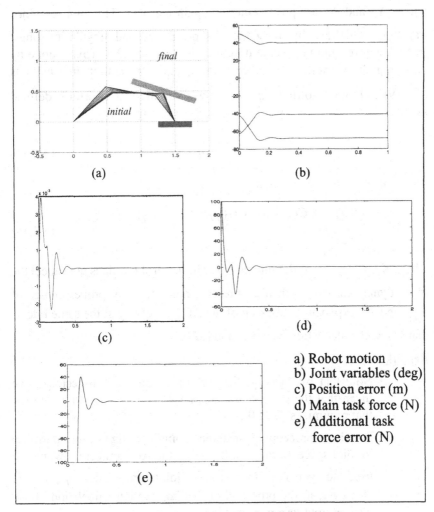

(a)

(b)

(c)

(d)

a) Robot motion
b) Joint variables (deg)
c) Position error (m)
d) Main task force (N)
e) Additional task
   force error (N)

(e)

Figure 4.13 Force-controlled additional task

$$\dot{X}^r = \dot{X}^t - \Lambda_x (X - X^t) \qquad (a)$$

$$\ddot{X}^r = \ddot{X}^t - \Lambda_x (\dot{X} - \dot{X}^t) \qquad (b)$$

$$\dot{Z}^r = \dot{Z}^t - \Lambda_z (Z - Z^t) \qquad (c)$$

$$\ddot{Z}^r = \ddot{Z}^t - \Lambda_z (\dot{Z} - \dot{Z}^t) \qquad (d)$$

$$(4.3.12)$$

where Λ_x and Λ_z are positive-definite gain matrices. The joint reference trajectory is defined by using the task prioritized and singularity robust redundancy resolution scheme described in Section 4.3.1. This is done by replacing the Cartesian reference velocity and acceleration in equations (2.3.19) and (4.3.4) to find $\dot{q}^r, \ddot{q}^r$. Now a virtual velocity error is defined as:

$$s = \dot{q} - \dot{q}^r \qquad (4.3.13)$$

The control law is then given by:

$$\tau = H(q)\ddot{q}^r + C(q, \dot{q})\dot{q}^r + G(q) + f(\dot{q}) + K_D s - J_e^T F_x^e$$
$$- J_{c1}^T F_z^e \qquad (4.3.14)$$

where K_D is a positive-definite matrix. This control law does not cancel the robot dynamics. However, it ensures asymptotic, or by proper choice of K_D, and Λ, exponential tracking of the ACT trajectory at the same rate as that of exact cancellation (see [81] and [82]).

Remarks:

- Note that by "asymptotic tracking of the ACT trajectory", we mean that the control law guarantees convergence to a solution that minimizes (2.3.20).

- The above procedure is different from the design of the controller in joint space, because in the latter, the ACT trajectory would be used to generate the desired joint trajectories $q^d, \dot{q}^d, \ddot{q}^d$. However, in the proposed algorithm, explicit calculation of the desired joint values is avoided.

- The use of the controller proposed in this section has two major advantages over the inverse dynamics (or computed torque) method which is used in Section 4.3.2:

    (i) It controls self-motion because both velocity and acceleration information are used; the computed torque method requires only a commanded acceleration trajectory.

    (ii) The formulation of this algorithm is similar to a non-adaptive version of the approach of Slotine and Li [81].

Therefore, to deal with inaccurate dynamic parameters, an adaptive implementation of this algorithm can be developed without major modifications to the inner loop which is the subject of Section 4.3.4 .

4.3.3.3 Simulation Results for a 3-DOF Planar Arm

The setup for the constrained compliant motion control is shown in Figure 4.6. A general block diagram of the simulation is shown in Figure 4.14.

<u>Obstacle avoidance with self-motion stabilization</u>

In this simulation, the end-effector is initially at rest and touches the constraint surface ($f=0$) at the point (1.5,0). The main task consists of keeping the position in the X direction constant, while exerting a desired -100 N in the Y direction. There is also a moving object enclosed in a circle in the workspace. The additional task consists of using the redundant degree of freedom to avoid this object. The simulation is carried out to compare the method proposed in Section 4.3.2 and the method proposed in this section.

As we can see in the plot of the joint velocities (Figure 4.15c, Figure 4.16c), there is a movement for a short period at the beginning to achieve the desired force - the end-effector moves in the Y direction to penetrate the surface. The manipulator remains stationary until the object is close enough to the arm. The obstacle avoidance task becomes active and makes the manipulator move in the null space of the Jacobian matrix to avoid collision

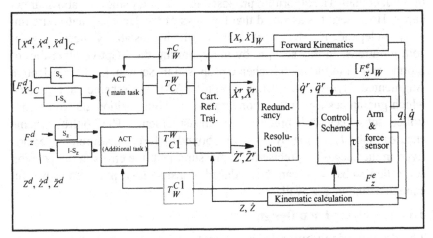

Figure 4.14 General block diagram of the AHIC scheme

while satisfying the main task. The two algorithms perform in the same way up to the point that the object clears the arm. From that point onwards, the algorithm in Section 4.3.2 is unable to control the null space components of the joint velocities and causes self-motion (Figure 4.15b). However, the proposed algorithm is successful in damping out these components and preventing self-motion.

4.3.4   Adaptive Augmented Hybrid Impedance Control

It has been shown that control methods that do not address uncertainties in a manipulator's dynamics may result in unstable motion in practice. This has led to considerable work on adaptive control of manipulators [59], [82]. Adaptive compliant control has also been addressed in recent years. Han et al. [27] have proposed an adaptive control scheme for constrained manipulators based on a nonlinear coordinate transformation; Lu and Meng [41] have proposed an adaptive impedance control scheme, and Niemeyer and Slotine [52] have discussed an application of the adaptive algorithm of Slotine and Li [81] to compliant motion control and redundant manipulators. However, application of the above algorithms to redundant manipulators introduces several problems. For instance, the algorithm in [27] requires definition of a nonlinear invertible transformation from joint space to a generalized task space. The algorithm in [41] is based on the Cartesian dynamic model of a manipulator and can be applied to the redundant case. However, no user defined additional tasks can be incorporated in the algorithm and redundancy is based on the generalized inertia-weighted inverse of the Jacobian. The algorithm proposed in [41] overcomes the above drawbacks. However, it is assumed that the rows of the Jacobian matrix are linearly independent. Hence, it may result in instability near singular configurations. In this section, by incorporating the adaptive algorithm of Slotine and Li in the AHIC scheme proposed in Section 4.3.3, an Adaptive Augmented Hybrid Impedance Control (AAHIC) scheme is presented which guarantees asymptotic convergence in both position and force controlled subspaces with precise force measurements. The control scheme ensures stability of the system with bounded force measurement errors. Even in the case of imprecise force measurement, the errors in the position controlled subspaces can be reduced considerably provided powerful enough actuators are available.

4.3.4.1   Outer-Loop Design

The design of the outer-loop is similar to that described in Section 4.3.3.1.

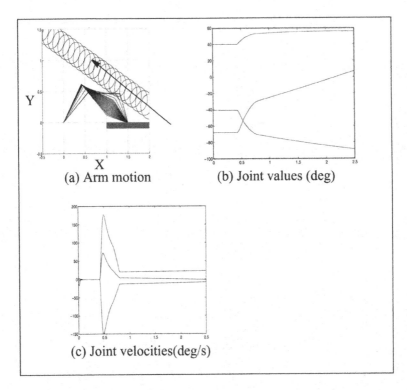

(a) Arm motion   (b) Joint values (deg)

(c) Joint velocities(deg/s)

Figure 4.15 Object avoidance without self-motion stabilization

4.3.4.2 Inner-Loop Design

The dynamics of a rigid manipulator are described by equation (4.3.8). The controller should be designed to calculate the torque input to equation (4.3.8), which ensures the tracking of the ACT trajectory in the presence of uncertainties in the manipulator's dynamic parameters.

It has been shown that for a suitably selected set of dynamic parameters, equation (4.3.8) can be written as:

$$H(q)\ddot{q}^r + C(q, \dot{q})\dot{q}^r + G(q) + f(\dot{q}) = Y(q, \dot{q}, \dot{q}^r, \ddot{q}^r)a \qquad (4.3.15)$$

where Y is the $n \times p$ regressor matrix and a is the $p \times 1$ vector of dynamic parameters. The matrix C is defined in such a way that $\dot{H} - 2C$ is a skew-symmetric matrix [81].

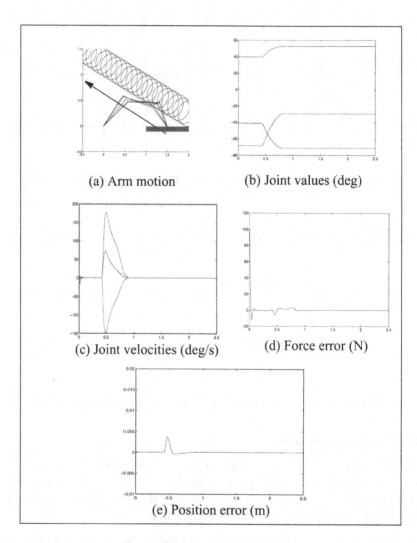

(a) Arm motion   (b) Joint values (deg)

(c) Joint velocities (deg/s)   (d) Force error (N)

(e) Position error (m)

Figure 4.16 Moving object avoidance with self-motion stabilization

Now an extension of the adaptive algorithm of Slotine and Li [81] is used to design the controller in order to ensure asymptotic tracking of the ACT trajectory. The procedure is as follows:

First, a Cartesian reference trajectory is defined for both the main and additional tasks (see equations (4.3.12)). Then, a virtual velocity error is defined (see (4.3.13)). The control law is then given by:

$$\tau = Y\hat{a} - K_D s - J_e^T \hat{F}_x^e - J_{c1}^T \hat{F}_z^e$$
$$= \hat{H}(q)\ddot{q}^r + \hat{C}(q,\dot{q})\dot{q}^r + \hat{G}(q) + \hat{f}(\dot{q}) - J_e^T \hat{F}_x^e - J_{c1}^T \hat{F}_z^e \qquad (4.3.16)$$

where $\hat{H}, \hat{C}, \hat{G}, \hat{f}, \hat{a}$ are calculated based on estimated values of $H, C, G,$ $f,$ and a respectively. $\hat{F}_x^e$ is the measured end-effector interaction force with the environment, K_D is a positive-definite matrix, and $s = \dot{q} - \dot{q}^r$. The last term on the right-hand side of the equation is only needed if another point of the manipulator (other than the end-effector) is in contact with the environment; F_z^e denotes the measured reaction force corresponding to a second constraint surface, and J_{c1} is the Jacobian of the contact point. We use the same Lyapunov candidate function as in [41]:

$$V(t) = \frac{1}{2}[s^T H s + \tilde{a}^T \Gamma \tilde{a}] \qquad (4.3.17)$$

where Γ is a constant positive-definite matrix and $\tilde{a} = a - \hat{a}$. Differentiating $V(t)$ along the trajectory of the system (4.3.8) leads to

$$\dot{V}(t) = -s^T K_D s + s^T Y \tilde{a} + s^T J_e^T \tilde{F}_x^e + s^T J_{c1}^T \tilde{F}_z^e \qquad (4.3.18)$$

where $\tilde{F} = F - \hat{F}$ denotes force measurement error. This suggests that the adaptation law should be selected as:

$$\hat{a} = -\Gamma Y^T s \qquad (4.3.19)$$

With this adaptation law, equation (4.3.18) leads to:

$$\dot{V}(t) = -s^T K_D s + s^T (J_e^T \tilde{F}_x^e + J_{c1}^T \tilde{F}_z^e)$$
$$\leq -\underline{k}_D \|s\|^2 + \|s\|(\|J_e\|\|\tilde{F}_x^e\| + \|J_{c1}\|\|\tilde{F}_z^e\|) \qquad (4.3.20)$$

and

$$\dot{V}(t) \leq -\underline{k}_D \|s\|^2 + \delta\|s\| \qquad (4.3.21)$$

where $\underline{k}_D$ is the minimum eigenvalue value of the matrix K_D, and δ satisfies the following inequality:

$$\left\|J_e\right\|\left\|\tilde{F}_x^e\right\| + \left\|J_{c1}\right\|\left\|\tilde{F}_z^e\right\| \le \delta \tag{4.3.22}$$

We also assume that $\left\|J_e\right\| \le \alpha$ and $\left\|J_{c1}\right\| \le \beta$. Now, we consider two different cases: precise and imprecise force measurements.

Precise force measurements $\tilde{F} = 0$

In this case, inequality (4.3.21) reduces to

$$\dot{V}(t) \le -k_D\|s\|^2 \tag{4.3.23}$$

which implies $a, s \in L_\infty^n$ or boundedness of a and s. Moreover, it can be shown that

$$\|s\|^2 dt \le \frac{-1}{k_D} dV(t) \tag{a}$$

$$\int_0^\infty \|s\|^2 dt \le \frac{-1}{k_D} \int_0^\infty dV(t) = \frac{1}{k_D}(V(0) - V(\infty)) \tag{b}$$

(4.3.24)

which implies that $s \in L_2^n$ and consequently $J_e s, J_c s \in L_2^n$. In order to establish a link between S and the tracking errors of ACT trajectories, we assume that the tracking errors of the damped least-squares solution (2.3.19) are negligible. Therefore, multiplying both sides of equation (4.3.13) by the augmented Jacobian, leads to

$$J_e s = \dot{e}_x + \Lambda_x e_x \tag{a}$$

$$J_c s = \dot{e}_z + \Lambda_z e_z \tag{b}$$

(4.3.25)

where

$$e_x = X - X^t \qquad e_z = Z - Z^t \tag{4.3.26}$$

The equations in (4.3.25) represent strictly proper, asymptotically stable linear time-invariant systems with inputs $J_e s, J_c s \in L_2^n$ which imply exact tracking and asymptotic convergence of the trajectories X and Z to the ACT trajectories [54], [59].

Imprecise Force Measurements $\tilde{F} \neq 0$ (Robustness Issue)

To take into account the robustness issue, we consider the effects of imprecise force measurements. It is obvious that error in force measurements directly affects the tracking performance in the force controlled subspaces of the main and additional tasks. However, we can show boundedness of the closed-loop trajectories. Moreover, the upper-bound on the error in the position-controlled subspaces can be reduced.

In this case, the time derivative of the Lyapunov candidate function satisfies

$$\dot{V}(t) \leq -\underline{k}_D \|s\|^2 + \delta \|s\| \tag{4.3.27}$$

As in [41], we can state that $\dot{V}(t)$ is not guaranteed to be negative semi-definite with an arbitrary value of $\underline{k}_D$ and a large δ for small values of $\|s\|$.

However, positive $\dot{V}(t)$ implies increasing V and subsequently $\|s\|$, which eventually makes $\dot{V}(t)$ negative. Therefore, s remains bounded and converges to a residual set. For a fixed value of $\underline{k}_D$, the lower bound on s is determined by $\delta / \underline{k}_D$ and can be reduced by selecting a larger value of $\underline{k}_D$. Note that larger $\underline{k}_D$ increases the control effort and may saturate the actuators. Using equations (4.3.24) and boundedness of s, we can conclude boundedness of e_x and e_Z.

Remark: Dawson and Qu [17] have proposed a modification to the control law given in (4.3.16) by adding a term $-K_\delta \mathrm{sgn}(s)$ to the right hand side with $K_\delta > \delta$. This eventually leads to the same inequality for $\dot{V}(t)$ as in (4.3.23) which implies asymptotic convergence of the errors. However, the control law proposed in [17] is discontinuous in terms of s and may excite unmodeled high-frequency dynamics.

4.3.4.3   Simulation Results for a 3-DOF Planar Arm

The setup for constrained compliant motion control is shown in Figure 4.6. A general block diagram of the simulation is shown in Figure 4.14.

Tool Orientation Control

In this simulation the additional task is defined as the control of the orientation of a tool attached to the end-effector. In this case, the desired value is specified as $q_3 = -85°$. The end-effector is initially at the point ($X=1$, $Y=1$) (Figures 4.17a, c) in touch with the surface (zero interaction force). Figures 4.17a, b show that without activating the additional task, there is no restriction on joint three. However, by activating the additional task (Figures 4.17c, d), the tool orientation is maintained at the desired value. Figures 4.18a, b show the errors in the position- and force-controlled subspaces which practically converge to zero. The dynamic parameter estimates and the velocity error are shown in Figures 4.18d, e.

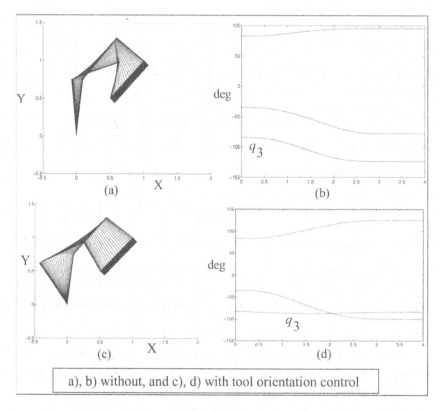

a), b) without, and c), d) with tool orientation control

Figure 4.17 Adaptive AHIC: Arm configuration and joint values

In order to study the effects of imprecise force measurements, the actual interaction force is augmented by a random noise uniformly distributed in the interval (-15N,15N). As we can see in Figure 4.19b, the error in the force controlled direction increases significantly as expected. The reason is that the controller in the force-controlled direction is based on force measurements and any error in this respect, directly affects the force error, e.g., the interval between 2 to 3 seconds. However, the error in the position-controlled direction (Figure 4.19a) remains practically unchanged from that of the previous simulation (Figure 4.18a), showing the robustness of the algorithm to force measurement error.

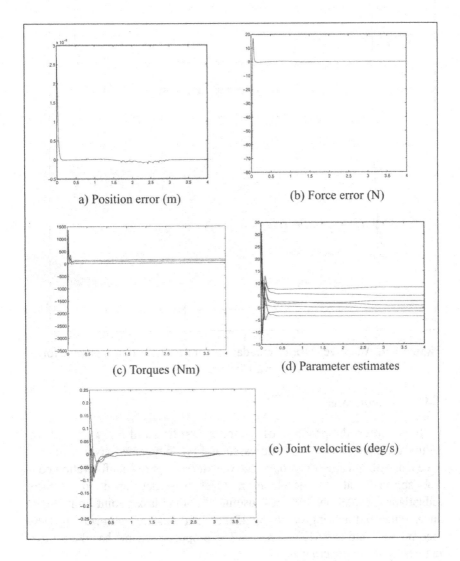

a) Position error (m)

(b) Force error (N)

(c) Torques (Nm)

(d) Parameter estimates

(e) Joint velocities (deg/s)

Figure 4.18 Adaptive AHIC with tool orientation control

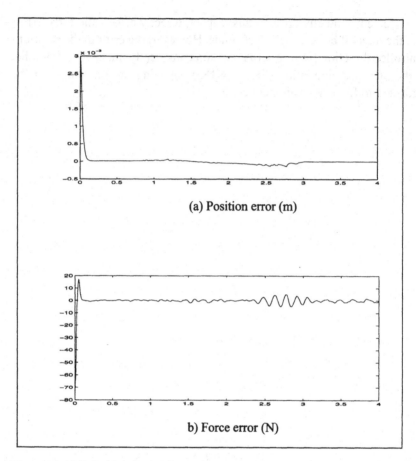

(a) Position error (m)

b) Force error (N)

Figure 4.19 Adaptive Hybrid Impedance Control: Effect of imprecise force
measurement

4.4   Conclusions

In this chapter, the problem of compliant motion and force control for
redundant manipulators was addressed and an Augmented Hybrid Impedance Control Scheme was proposed. An extension of the configuration control approach at the acceleration level was developed to perform
redundancy resolution. The most useful additional tasks: Joint limit avoidance, static and moving object avoidance, and posture optimization, were
incorporated into the AHIC scheme. The proposed scheme has the following desirable characteristics:

- Different additional tasks can be easily incorporated into the AHIC scheme without modifying the scheme and the control law.

- The additional task(s) can be included in the force-controlled subspace of the augmented task. Therefore, it is possible to have a multiple-point force control scheme.

- Task priority and singularity robustness formulation of the AHIC scheme relax the restrictive assumption of having a non-singular augmented Jacobian.

A modified AHIC scheme was proposed in this chapter that gives a solution to the undesirable self-motion problem which exists in most dynamic control schemes developed for redundant manipulators. An Adaptive Augmented Hybrid Impedance Control (AAHIC) scheme was described which guarantees asymptotic convergence in both position- and force-controlled subspaces with precise force measurements. The control scheme also ensures stability of the system in the presence of bounded force measurement errors. Even in the case of imprecise force measurements, the errors in the position controlled subspaces can be reduced considerably. The performance of the proposed AHIC schemes was illustrated for a 3-DOF planar arm. In the next chapter, we will extend the AHIC scheme to the 3-D workspace of REDIESTRO, a 7-DOF experimental robot.

5 Augmented Hybrid Impedance Control for a 7-DOF Redundant Manipulator

5.1 Introduction

In Chapter 4, the AHIC scheme was developed and verified by simulation on a 3-DOF planar arm. In this chapter the extension of the AHIC scheme to the 3-D workspace of REDIESTRO, a 7-DOF experimental manipulator, is described. Figure 5.1 shows a simplified block diagram of the AHIC controller. Considering that the capabilities of the redundancy resolution scheme with respect to collision avoidance have already been fully demonstrated, in order to focus on the new issues related to Contact Force Control (CFC), the environment is assumed to be free of obstacles.

The complexity of the required algorithms and constraints on the amount of computational power available have resulted in an algorithm development procedure which incorporates a high level of optimization. At the same time, the following issues which were not studied in the 2-D workspace need to be tackled in extending the schemes to a 3-D workspace:

- Extension of the AHIC scheme for orientation and torque

- Control of self-motion as a result of resolving redundancy at the acceleration level for the AHIC scheme represented in Section 4.3.2

- Robustness with respect to higher-order unmodelled dynamics (joint flexibility), uncertainties in manipulator dynamic parameters, and friction model.

5.2 Algorithm Extension

In this section, the different modules involved in the AHIC scheme are described. The focus is on describing the required algorithms without getting involved in the specific way in which the modules are implemented.

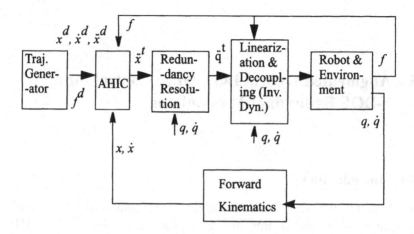

Figure 5.1 Simplified block diagram of the AHIC controller

5.2.1   Task Planner and Trajectory Generator (TG)

The robot's task can be specified using a Pre-Programmed Task File. Each line indicates the desired position and orientation to be reached at the end of that segment, the hybrid task specification, and the desired impedance and force (if applicable) for each of the 6 DOFs.

In the absence of obstacles, the robot path will consist of straight lines connecting the desired position/orientation at each segment. The TG module generates a continuous path between the via points. The TG implemented to test the AHIC scheme generates a fifth-order polynomial trajectory which gives continuous position, velocity, and acceleration profiles with zero jerk (rate of change of acceleration) at the beginning and the end of the motion.

5.2.2   AHIC module

Figure 5.2 shows the location of the different frames used by the AHIC module. The description of the environment is specified in a configuration file. As an example, for a surface-cleaning task, it is required to specify the location and orientation of a fixed frame $\{C\}$ with respect to the world frame. In this case, the robot's base frame $\{R_1\}$ is selected as the world frame. The tool frame $\{T\}$ is attached to the last link. Depending on the type of the tool, the user specifies the location and orientation of this frame in the last joint's local frame. The force sensor interface card also uses this information to locate the force sensor frame at $\{T\}$. The task frame $\{C_i\}$ is located at the origin of the frame $\{T\}$. However, the orientation of $\{C_i\}$ is dictated by $\{C\}$. Therefore, the frame $\{C_i\}$ moves with the tool while keeping the same orientation as the constant frame $\{C\}$.

The AHIC scheme, as implemented for the 2-D workspace, generates an Augmented Cartesian Target Acceleration *(ACTA)* for the end-effector (EE) position in real-time:

$$\ddot{X}^t = M^{d^{-1}}(-F^e + (I - S)F^d - B^d(\dot{X} - S\dot{X}^d) - K^d S(X - X^d)) \\ + S\ddot{X}^d \tag{5.2.1}$$

where M^d, B^d, K^d are diagonal matrices whose diagonal elements represent the desired mass, damping, and stiffness; S is a diagonal selection matrix which specifies the force- ($S_i = 0$) or position- ($S_i = 1$) controlled axis; F^d, F^e are the desired and interaction forces.

In order to keep the concept of splitting position and orientation control as described in Section 3.3.2 , the ACTA in the 3-D workspace will be generated separately for position/force-controlled and orientation/torque-controlled axes:

$$\ddot{P}^t(t) = M_p^{d^{-1}}(-F^e + (I - S_p)F^d - B_p^d(\dot{P} - S_p\dot{P}^d) \\ - K_P^d S_p(P - P^d)) + S_p\ddot{P}^d \tag{5.2.2}$$

$$\dot{\omega}^t(t) = M_o^{d^{-1}}(-N^e + (I - S_o)N^d - B_o^d(\omega - S_o\omega^d) \\ - K_o^d S_o e_o) + S_o\dot{\omega}^d \tag{5.2.3}$$

where the subscripts p and o indicate that the corresponding variables are specified for position/force-controlled and orientation/torque-controlled subspaces respectively. The superscript d denotes the desired values. The vector $P(3 \times 1)$ and its derivatives are the position, velocity, and acceleration of the origin of $\{T\}$ expressed in frame $\{C\}$; F^d and F^e are the desired and interaction forces expressed in $\{C\}$; $S_p(3 \times 3)$ is the selection matrix used to indicate that a {C} frame axis is force- or position-controlled; ω, $\dot{\omega}$ are the angular velocity and acceleration of the {T} frame expressed in {C_i} ; e_o is the orientation error vector (see Section 3.3.2.2); N^d, N^e are the desired and interaction torques in frame {C_i} ; and M^d, B^d, K^d are diagonal matrices whose diagonal elements represent the desired mass, damping, and stiffness.

Equation (5.2.2) is resolved in frame {C} while Equation (5.2.3) is resolved in frame {C_i}. The frame {C_i} is a time-varying frame (in contrast to frame {C} which is a fixed frame) located at the origin of frame {T} and with same orientation as {C}.

All the inputs and outputs in equations (5.2.2) and (5.2.3) should be expressed in frames {C} and {C_i} respectively. In order to make the AHIC controller module self-contained, all the necessary conversions are implemented in this module.

The location of the origin of {C} in {R_1} ($^{R_1}P_C$) and the (3 × 3) rotation matrix $^{R_1}R_C$ are specified in a configuration file. It should be noted that the orientations of {C} and {C_i} in any arbitrary frame are the same.

5.2.3   Redundancy Resolution (RR) module

The RR module for the AHIC scheme should be implemented at the acceleration level. Assuming an obstacle-free workspace, the damped least-squares solution is given by:

$$\ddot{q}^t = A^{-1}b \qquad (5.2.4)$$

where

$$A = J_p^T W_p J_p + J_o^T W_p J_p + J_c^T W_c J_c + W_v$$
$$b = J_p^T W_p(\ddot{P}^t - \dot{J}_p\dot{q}) + J_p^T W_p(\dot{\omega}^t - \dot{J}_o\dot{q}) + J_c^T W_c \ddot{Z}^t$$

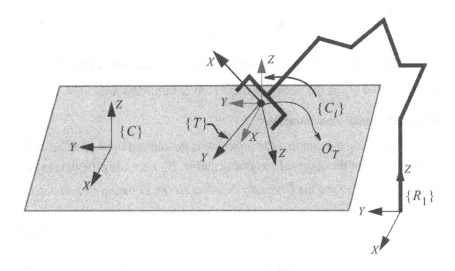

Figure 5.2 Different frames involved in the hybrid task specification

J_p and J_o are the Jacobian matrices projecting the joint rates to linear and angular velocities of frame {T}. The Jacobian matrices and the two vectors ($\dot{J}_p\dot{q}, \dot{J}_o\dot{q}$) are calculated by the forward kinematics module. The matrices W_p, W_o, W_v are the diagonal weighting matrices that assign priority between position/force tracking, orientation/torque tracking and singularity avoidance (in the case of conflicts between these tasks), these matrices are specified by the user in a configuration file. A complete study that demonstrates the effects of the weighting matrices is given in Section 3.3.2.3 . The vectors $\ddot{P}^t, \dot{\omega}^t$ are the target linear and angular accelerations of frame {T} expressed in the robot's base frame. These vectors are calculated by the AHIC module. Because the quantities are expressed in the same frame, no coordinate transformation is needed. Note that at this stage, the additional task that is incorporated into the system is joint limit avoidance. For the joint limit avoidance task, the terms $J_c^T W_c J_c$ and $J_c^T W_c$ reduce to W_c (see Section 2.4.1.3). The target acceleration for the *ith* joint in the case of violation of soft-joint limits is defined by:

$$\dot{Z}_i^t = -K_{v_i}\dot{q}_i - K_{p_i}(q_i - q_{m_i}) \qquad (5.2.5)$$

where K_p and K_v are positive-definite proportional and derivative gain matrices, and q_m is the vector of maximum or minimum joint limits.

Computational considerations:

Considering the fact that the matrix A is guaranteed to be positive definite (because of the diagonal weighting matrix W_v), a more efficient way to solve (5.2.5) is to use the Cholesky decomposition. Equation (5.2.4) can be written in the form

$$Ax = b \qquad (5.2.6)$$

where $x = \ddot{q}^t$. The Cholesky decomposition of A is given [93] by: $A = LL^T$, where L is a lower-triangular matrix. This reduces to solving an upper and an lower-triangular system of linear equations:

$$Ly = b, \qquad L^T x = y \qquad (5.2.7)$$

5.2.4   Forward Kinematics

This module calculates the position and orientation of frame {T}, the linear and angular velocities of {T}, and also the Jacobian matrices relating the linear and angular velocities of {T} to the joint rates. These quantities are expressed in the robot's base frame.

- **Tool frame Information:** It is only necessary to specify the information to locate frame {T} in frame {7}. Therefore, $Twist(\alpha_7)$, $Length(a_7)$, $Offset(d_7)$, are specified in a configuration file which results in:

$$
{}^7T_T = \begin{bmatrix}
1 & 0 & 0 & a_7 \\
0 & \cos(\alpha_7) & -\sin(\alpha_7) & 0 \\
0 & \sin(\alpha_7) & \cos(\alpha_7) & d_7\sin(\alpha_7) \\
0 & 0 & 0 & 1
\end{bmatrix}
$$

- **Calculation of** $\dot{J}_p\dot{q}, \dot{J}_o\dot{q}$: Calculation of two new vectors ($\dot{J}_p\dot{q}, \dot{J}_o\dot{q}$) which are required by the RR module (because of resolving redundancy at the acceleration level) are added to the forward kinematics module. The forward kinematics function at the acceleration level is defined by:

$$\ddot{X} = J\ddot{q} + \dot{J}\dot{q} \tag{5.2.8}$$

which yields

$$\ddot{X}\big|_{\ddot{q}=0} = \dot{J}\dot{q} \tag{5.2.9}$$

This suggests that the following recursive algorithm, which calculates the linear and angular accelerations of the frame {T}, can be used to calculate the vectors ($\dot{J}_p\dot{q}, \dot{J}_o\dot{q}$).

for $i = 1...n+1$

$${}^{i}\omega_{i-1} = {}^{i}R_{i-1}{}^{i-1}\omega_{i-1}$$

$${}^{i}\omega_{i} = {}^{i}\omega_{i-1} + \dot{q}_{i}z_{i}$$

$${}^{i}\dot{\omega}_{i} = {}^{i}R_{i-1}{}^{i-1}\dot{\omega}_{i-1} + {}^{i}\omega_{i-1} \times \dot{q}_{i}z_{i}$$

$${}^{i}\dot{v}_{i} = {}^{i}R_{i-1}({}^{i-1}\dot{v}_{i-1} + {}^{i-1}\dot{\omega}_{i-1} \times {}^{i-1}P_{i} + {}^{i-1}\omega_{i-1}$$

$$\times ({}^{i-1}\omega_{i-1} \times {}^{i-1}P_{i}))$$

with initial values:

$${}^{0}\omega_{0} = [0, 0, 0]^{T}, {}^{0}\dot{v}_{0} = [0, 0, 0]^{T} \tag{5.2.10}$$

Note that the frames {8} and {T} are the same, and also, the frame {0} is located at the robot's base frame {R1}. Now, equation (5.2.9) results in:

$$\dot{J}_p\dot{q} = {}^{0}R_{T}{}^{8}\dot{v}_{8}, \qquad \dot{J}_o\dot{q} = {}^{0}R_{T}{}^{8}\dot{\omega}_{8} \tag{5.2.11}$$

5.2.5   Linear Decoupling (Inverse Dynamics) Controller

The equation of motion of a 7-DOF manipulator, considering interaction forces/torques with its environment, is given by

$$M(q)\ddot{q} + H(q, \dot{q}) + G(q) + f(q, \dot{q}) = \tau - J^T F \qquad (5.2.12)$$

where M is the (7×7) symmetric positive-definite inertia matrix of the manipulator in joint space; H is the (7×1) vector of centripetal and Coriolis torques, G is the (7×1) gravity vector, F is the (6×1) vector of interaction forces/torques exerted by the robot on the environment at the operating point (origin of the tool frame), J is the (6×7) Jacobian matrix relating the linear and angular velocities of the tool frame to joint rates, f is the (7×1) joint friction vector, and τ is the (7×1) vector of applied torques at the actuators.

The torque that is required to linearize and decouple the nonlinear equation (5.2.12) is given by:

$$t_{LD} = \tau_1 + \tau_2 \qquad (5.2.13)$$

where

$$\tau_1 = \hat{M}(q)\ddot{q} + \hat{H}(q, \dot{q}) + \hat{G}(q) + J^T \hat{F}$$
$$= InvDyn(q, \dot{q}, \ddot{q}^t, \hat{F}) \qquad (5.2.14)$$

and

$$\tau_2 = \hat{f}(q, \dot{q}) \qquad (5.2.15)$$

where $\wedge$ denotes the estimated values. The optimized *InvDyn* function as well as the closed-form representations of M, H, G are developed in C using the Robot Dynamics Modeling (RDM) software [78].

5.3   Testing and Verification

In the simulation developed for the purpose of verifying the integration of the controller, the inverse dynamics and the model of the arm are replaced by double integrators, i.e., we assume perfect knowledge of the manipulator dynamics. However, the model of the environment is still present. The environment is modeled by a linear spring.

To verify and test the integration of the controller modules, we recall that if the AHIC scheme is successful, the manipulator acts as a desired impedance in each of the 6 DOF's of the {C} frame. Figure 5.3 shows the desired impedance in position-controlled and force-controlled axes respectively. In order to verify the operation of the AHIC scheme, two simple one-dimensional simulations for the position and force controlled axes were used (see Figure 5.4).

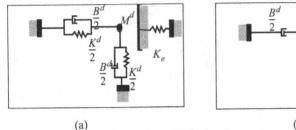 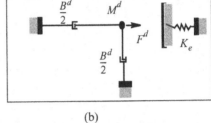

(a)   (b)

Figure 5.3 Desired impedance a) position controlled axis, and b) force-controlled axis

To check the correct operation of the controller in position-controlled directions, all axes were specified to be in position-control mode. A 1 N symmetric step force (in all three X, Y, and Z dimensions of the {C} frame) was applied to both systems. The desired impedance values can be selected arbitrarily at this stage, because we only need to compare the responses of the two systems. The impedance values used for this test in all 6 DOF's of the {C} frame were

$$M^d = 257kg, B^d = 1100\frac{Ns}{m}, K^d = 11000\frac{N}{m}$$

$$\Rightarrow \xi = 0.32, \omega_n = 6.54$$

Figure 5.5 shows the plots of the changes in the position of the origin of the frame {T} along X and Y axes of the {C} frame. The same test was performed for the force-controlled direction with the following values:

$$M^d = 257kg, B^d = 1100\frac{Ns}{m}, K_e = 11000\frac{N}{m}, F^d = 20N$$

Figure 5.6 compares the force history of the AHIC after contacting the surface with that of the pure-impedance simulation in Figure 5.4b. As one can see, the response of the AHIC simulation is very close to that of the pure impedance simulation. The possible sources of the small discrepancies are as follows:

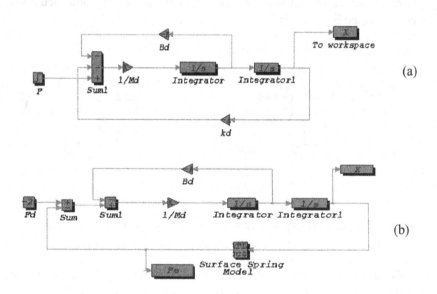

(a)

(b)

Figure 5.4 *Simulink* one-dimensional simulation of the desired impedance
a) position-controlled axis, b) force controlled axis.

- As mentioned in Section 3.3.2.3 , the presence of the singularity robustness term (W_v) introduces some error.

- The simulation of the AHIC scheme is a discrete-time simulation with a trapezoidal integration routine written in C, in contrast to the imped-ance simulation which is run in continuous-time mode.

- In the AHIC simulation some delays are added to break the "alge-braic-loops". These are not present in the ideal impedance system simula-tion shown in Figure 5.4.

Note that the test results up to this point show the correct integration of different modules. Detailed study and analysis of the performance are described in the next section.

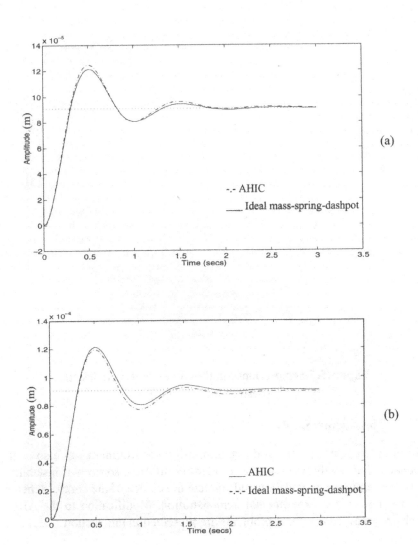

Figure 5.5 Step response in position-controlled directions - Position of the origin of {T} (expressed in {C}) in response to a step force of 1N. a) X axis, b) Y axis

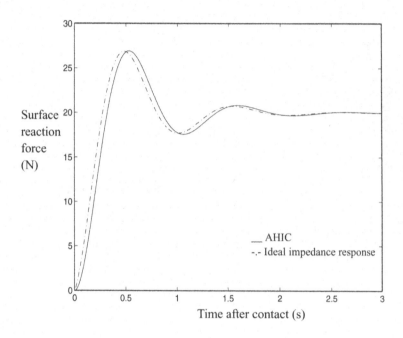

Figure 5.6 Step response in the force-controlled direction
(desired force = 20N)

5.4   Simulation Study

In order to perform this study, a simulation environment has also been created. This study will allow us to identify different sources of instability and performance degradation and finalize the choice of the control scheme to be used in the experimental demonstration. Modification to the AHIC scheme to overcome these problems are presented in this section.

5.4.1   Description of the simulation environment

The control modules in the simulation are described in Section 5.2. The robot model (Figure 5.7) has been developed using the RDM software [78]. It models REDIESTRO and its hardware accessories and covers the following main features:

- Optimized forward dynamics module of the arm;
- Joint friction including stiction, viscous, and Coulomb friction;

- Digitization effects of the A/D converters and encoders;
- Saturation of the actuators and current amplifiers

It also provides some additional features:

- Optimized closed-form representations of the inertia matrix, Coriolis, and gravity vectors;
- Effect of external forces;
- Surface and force-sensor models.

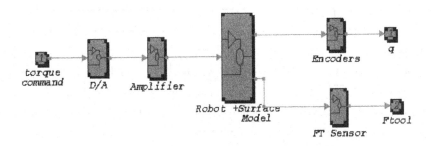

Figure 5.7 Simulation model of REDIESTRO with the addition of a force sensor and surface models.

In order that the simulation be as close as possible to reality, the simulation is implemented in a mixed discrete and continuous mode. The robot and the surface models use a continuous simulation (Runge-Kutta 5th-order integration), and all other modules are discrete modules with a sampling frequency of 200 Hz.

The joint angles and the interaction forces are transferred via an ethernet network to another SGI workstation which runs the Multi-Robot Simulation (MRS) graphical software [77], [9], [10] for online 3-D graphics rendering of the movement of the arm.

5.4.2 Description of the sources of performance degradation

In this section by using different simulations and hardware experiments, we determine the sources of degradation in the performance and, in the extreme case, instability, and suggest modifications that can deal with these problems.

Figure 5.8 shows a simplified block diagram of the simulation of the AHIC scheme. The major sources of performance degradation and instability are as follows:

- Kinematic instability due to resolving redundancy at the acceleration level.

- Performance degradation due to the model-based part of the controller.

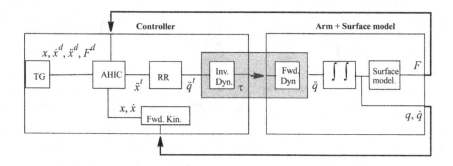

Figure 5.8 Simplified block diagram of the AHIC controller simulation.

In the following sections, these problems will first be demonstrated using simulation and/or hardware experiments. Then, the required modifications to the AHIC scheme will be described.

5.4.2.1 Kinematic instability due to resolving redundancy at the acceleration level

In order to focus on this specific problem, we assume that the inverse dynamics part of the controller perfectly decouples the manipulator's dynamics, so that, the arm model can be replaced by a double integrator. It was previously noted (see Section 4.3.3) that resolving redundancy at the acceleration level has the drawback that self-motions (joint motions that do not induce any movement in Cartesian space) of the arm are not controlled.

A simulation is performed with non-zero initial joint velocities. The robot is commanded to go from an initial position/orientation to a final position/orientation in 3 seconds and keep the same position/orientation thereafter (the desired velocity and acceleration are zero after 3 seconds).

As we can see in Figure 5.10, the robot tracks the trajectory very well. However, the controller is not able to damp out the self-motion component of the joint velocity after reaching the final point.

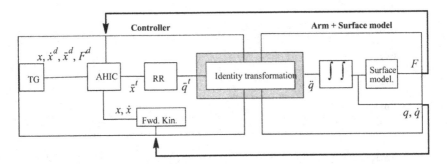

Figure 5.9 Simplified block diagram of the simulation used in kinematic instability analysis

The following solution can be used:

• Reducing the dimension of the self-motion manifold to zero by specifying additional tasks, e.g. freezing or controlling the value of one of the joints

• Using an improved redundancy resolution scheme at the acceleration level in order to achieve self-motion stabilization [32].

• Modifying the AHIC scheme in order to be able to use redundancy resolution at the velocity level (see Section 4.3.3).

Freezing or controlling the position of one of the joints, is not a preferable option, because that eliminates a desirable redundant degree-of-freedom which otherwise could be used to fulfill additional tasks. The solutions given for improved redundancy resolution at the acceleration level are computationally more expensive, because they require the explicit calculation of the derivative of the Jacobian matrix.

The AHIC scheme with self-motion stabilization, proposed in Section 4.3.3 , achieves its goal by modifying AHIC in order to use redundancy resolution at the velocity level. However, the model-based part of the controller (inner-loop) is much more complicated than the computed torque algorithm. The former requires tracking of a reference joint velocity - see equation (4.3.14). The key idea to solve this problem is to control the velocity. We propose the following to control the velocity:

$$\ddot{q} + \lambda \dot{q} = 0 \qquad (5.4.1)$$

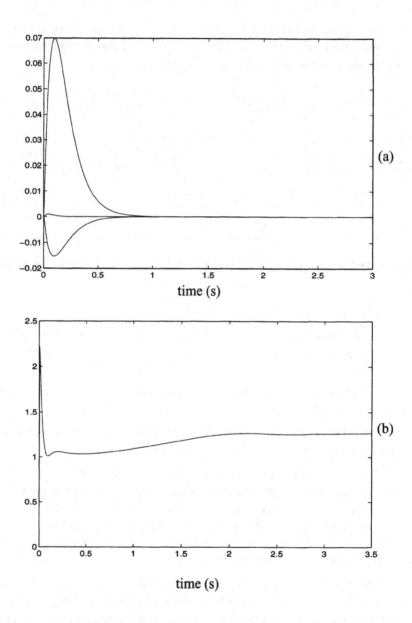

Figure 5.10 Simulation results with non-zero initial velocity: (a) position error (m); (b) norm of the joint velocities (rad/s).

This suggests a modification of the cost function in (4.3.2) to

$$L = \ddot{E}^T W \ddot{E} + (\ddot{q} + \lambda \dot{q})^T W_v (\ddot{q} + \lambda \dot{q}) \qquad (5.4.2)$$

The damped least-squares solution for the new cost function is given by:

$$\ddot{q} = (J^T W J + W_v)^{-1} (J^T W (\ddot{X}^t - \dot{J}\dot{q}) - W_v \lambda \dot{q}) \qquad (5.4.3)$$

which in fact penalizes non-zero velocities. To verify the performance of the modified redundancy resolution scheme, a simulation was performed. In order to verify the performance in the worst case, the final position/orientation was selected such that it makes the robot's posture approach a singular configuration. This, in fact, induces a high null-space component on the joint velocities. Again the robot was commanded to go from an initial position to a final position in 3 seconds. The robot should reach its final position in Cartesian space in 3 seconds. However, there is a large null-space component of the joint velocities that remains uncontrolled when $\lambda = 0$. Increasing the value of λ damps out these components (Figure 5.11).

In order to study the effect of λ on tracking error, another set of simulations was performed. shows the results of these simulations. As in the previous simulations the desired position is reached after 3 seconds. For $\lambda = 5$, the velocity fades away with large oscillations. With $\lambda = 50$ the velocity fades away with no overshoot. However, there are larger tracking errors. A choice of $\lambda = 10$ gives the best result considering both tracking and velocity damping. Based on our experience a value of λ between 7.5 and 12.5 was found to be suitable for most cases.

5.4.2.2 Performance degradation due to the model-based part of the controller

In order to focus on this specific problem, first let us consider the simpler case of a Linearized-Decoupled Proportional-Derivative (LDPD) joint-space controller as shown in Figure 5.13.

The model based part of the controller decouples and linearizes the manipulator's dynamics if both the model and the parameters used in the controller are perfect. However, in reality there are different sources of parameter and model mismatch. Some of the major sources of performance degradation of a model-based controller are listed below:

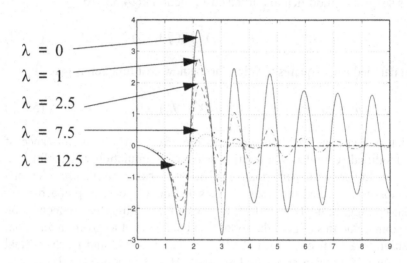

Figure 5.11 Simulation result for the modified RR scheme - Joint 2 velocity (rad/s).

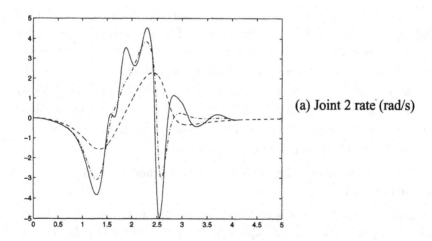

(a) Joint 2 rate (rad/s)

Figure 5.12 Comparison between different values of damping factors in the RR module.

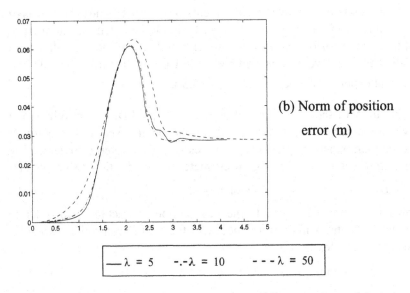

(b) Norm of position error (m)

Figure 5.12 (contd.) Comparison between different values of damping factors in the RR module.

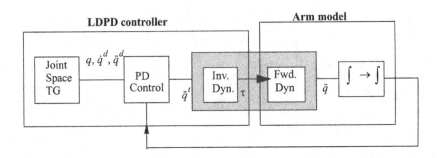

Figure 5.13 Block diagram of the LDPD controller

- Friction compensation (model & parameters)
- Unmodeled dynamics (e.g. joint flexibility)
- Imprecise dynamic and kinematic parameters
- Initial joint offsets

Simulations and hardware experiments were used to study the effects of these sources on tracking performance. It should be noted that in order to distinguish the performance of the model-based part of the controller from the PD part, we have not selected high gain values in the following simulations and experiments ($K_p = 10, K_v = 6.3$).

Figure 5.14 shows the simulation results of the LDPD controller (Joint 2) when the same friction model and parameters are used in the controller and the manipulator model. The errors essentially converge to zero. The simulation was repeated using an estimate of joint friction values greater that those used in the manipulator model $\tilde{f}_{fric} = 1.3 f_{fric}$. The results shown in Figure 5.15 indicate the degradation in tracking. In order to achieve better tracking performance with the LDPD controller, two solutions are available:

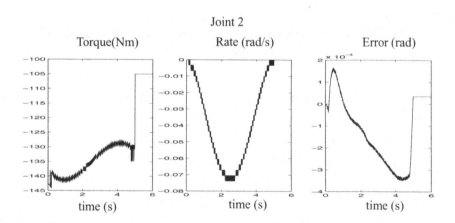

Figure 5.14 Simulation results for the LDPD controller using the same parameters and model in the inverse dynamic controller and the manipulator model ($K_p = 10, K_v = 6.3$)

- Increasing the feedback gains
- Better parameter identification

The first solution improves tracking. However, it decreases robustness because of the risk of exciting higher-order unmodeled dynamics, e.g. joint flexible modes. The second option also improves tracking. However, there is a limit to the accuracy level of identification for different manipulators.

Moreover, these parameters may change with time and the initially identified parameters may not be accurate after a certain time. This is important in space applications where, after launching the arm, periodic identification of parameters may not be feasible.

It should be noted that the AHIC scheme, shown in Figure 5.1, does not include any "control gains". The only gains in the AHIC scheme are the "impedance gains". The control gains can be selected arbitrarily based on the accuracy level of the modeling and tracking requirements. However, the criteria for selecting the impedances gains are dictated differently, e.g., by surface dynamics, stability considerations for the force control loop.

From the above statements, one can conclude that a way of improving the performance of the AHIC scheme is a combination of the following steps:

1-Adding a PD feedback loop in the AHIC scheme

2- "Better" parameter identification

- Refinement of the friction compensation module
- Fine tuning of friction coefficients
- Accurate home positioning

The following section describes the modification to the AHIC controller.

5.4.3 Modified AHIC Scheme

Section 5.4.2.2 indicated the problem associated with the model based controller using a simple example of a joint-space LDPD controller. Now, we study the same problem using the complete simulation of the AHIC scheme (see Figure 5.8). These simulations contain two segments: free motion and contact-motion. In the first segment, the tool frame is commanded to move from an initial position to a final position located on the constraintsurface in 3 seconds. The second segment consists of keeping the final position (along x and y) and orientation while exerting a 60 N force on the surface. Table 5-1 summarizes the values used in the simulations.

Figure 5.16 shows the force tracking when joint friction is not included in the joint models. As one can see, there is a small error between the joint target acceleration command ($\ddot{q}^t$) and the actual joint acceleration. This results in accurate tracking of the response of the desired impedance in the z direction.

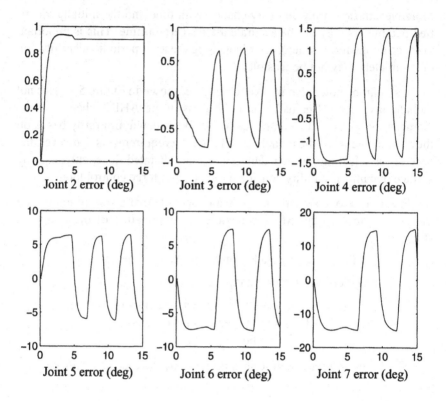

Figure 5.15 LDPD controller with inexact friction compensation

$$\tilde{f}_{fric} = 1.3 f_{fric}, K_p = 10, K_v = 6.3$$

Next, the simulation was repeated by including friction in the manipulator model. The friction compensation used the same model with exact parameters. Figure 5.17(b) shows the error between the target and the actual joint accelerations. Although both the friction model and the friction compensation module used the same model and parameters, the friction compensation used discretized data (e.g. joint velocities) while the friction model used continuous values. Figure 5.17(a) shows the tracking degradation resulting from this slight mismatch between the model and model-based controller. As mentioned earlier, the selection of the desired impedances is based on other criteria. Hence, they cannot be changed to deal with this problem.

Table 5-1 Desired values used in the AHIC simulation (z - axis)

seg.	S	M (kg)	B (Nsec/m)	K (N/m)	Fd (N)	Surface K (N/m)	Desired Eq. of motion
non contact	1	1	20	100	--	--	$M(\ddot{X}-\ddot{X}^d)+B(\dot{X}-\dot{X}^d)$ $+K(X-X^d)=-F_e$
contact	0	100	1000	--	60	10000	$M\ddot{X}+B\dot{X}-F^d=-F_e$

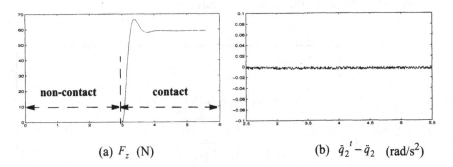

(a) F_z (N)   (b) $\ddot{q}_2^{\,t}-\ddot{q}_2$ (rad/s^2)

Figure 5.16 AHIC controller without joint friction

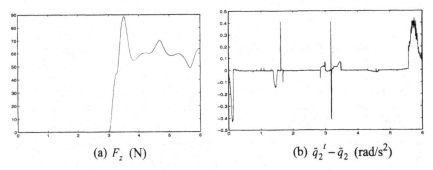

(a) F_z (N)   (b) $\ddot{q}_2^{\,t}-\ddot{q}_2$ (rad/s^2)

Figure 5.17 AHIC simulation with friction in the model and friction compensation in the controller (using the same friction parameters as in the model)

Adopting a similar scheme to that proposed in Section 4.3.3 , a solution to this problem is to add a PD feedback loop. Figure 5.18 shows the block diagram of the modified controller. The following modifications have been made:

- The Error Reference Controller (ERC) module which generates a Cartesian Reference Acceleration (CRA) has been added

- The position feedback which used to go to the AHIC module is now connected to ERC

- The complete target trajectory $(x^t, \dot{x}^t, \ddot{x}^t)$ is generated online using force sensor feedback

Figure 5.18 shows the new/modified modules which are shaded in gray. Table 5-2 summarizes the modified equations

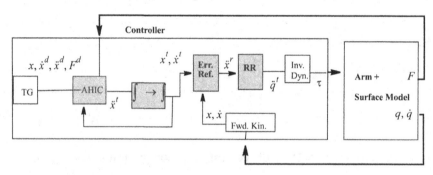

Figure 5.18 Simplified block diagram of the modified AHIC controller

Table 5-2 Summary of equations for new/modified modules

Module	Equation
AHIC	$\ddot{X}^t = M^{d^{-1}}(-F^e + (I-S)F^d - B^d(\dot{X}^t - S\dot{X}^d) - K^d S(X^t - X^d)) + S\ddot{X}^d$
ERC	$\ddot{X}^r = \ddot{X}^t + K_v(\dot{X}^t - \dot{X}) + K_p(X^t - X)$
RR	$\ddot{q}^t = (J^T W J + W_v)^{-1}(J^T W(\ddot{X}^r - \dot{J}\dot{q}) - \lambda W_v \dot{q})$

At this stage, another level of algorithm development was performed for the new/modified modules and functions. The complete simulation of the modified AHIC scheme was developed in the *Simulink* environment to study the performance of the modified scheme.

The simulations consist of 5 segments which are summarized in Table 5-3. The PD gains are chosen as $K_p = 100, K_v = 20$. The results of the original AHIC scheme are compared with the modified AHIC scheme. No joint friction compensation is performed to study the robustness of the algorithms.

Figure 5.19 shows the comparison between the force tracking performance of the AHIC scheme as shown in Figure 5.1, and that of the modified AHIC scheme. As one can see, even without performing friction compensation, the modified AHIC scheme is able to regulate the interaction force (with limited error). However, the original AHIC scheme is completely incapable of regulating the force. Note that force tracking can be greatly improved by selecting the appropriate impedance values (this will be explained in Section 6.2.1).

Table 5-3 Desired values used in the modified AHIC simulation (z - axis)

seg.	S	M (kg)	B (Nsec/m)	K (N/m)	Fd (N)	Surface K (N/m)	final_time (S)	Comment
1	1	1	20	100	--	--	3	non-contact
2	0	100	1000	--	60	10000	6	contact
3	0	100	1000	--	80	10000	7	contact
4	0	100	1000	--	40	10000	10	contact
5	0	100	1000	--	0	10000	13	contact

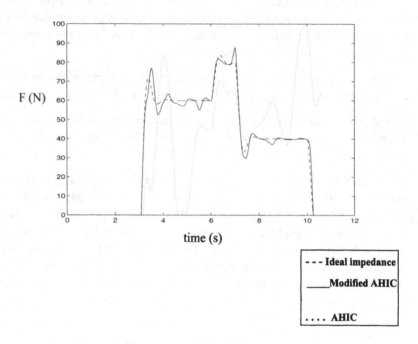

Figure 5.19 Comparison between the original AHIC scheme and the modified AHIC scheme (without friction compensation).

5.5 Conclusions

As indicated in the introduction, the objective of this chapter was to extend the AHIC scheme to the 3D workspace of a 7-DOF manipulator (REDIESTRO), to develop and test the AHIC software, and to demonstrate by simulation the performance of the proposed scheme. From the foregoing sections, the following conclusion can be drawn:

1. The conceptual framework presented for compliant force and motion control in the 2D workspace of a 3-DOF planar manipulator, is adequate to control a 7-DOF redundant manipulator working in a 3D workspace.

2. The algorithm extension for the AHIC scheme and the required modules have been successfully developed and implemented for REDIESTRO.

3. The software development of different modules has been successfully accomplished. The code has been optimized in order to achieve real-time implementation.

4. At this stage, only joint limit avoidance has been incorporated into the redundancy resolution module. The simulation results for joint limit avoidance provide confidence that other additional tasks such as obstacle avoidance can be incorporated without major difficulties.

5. The realistic dynamic simulation environment has enabled us to study issues such as performance degradation due to imprecise dynamic modelling and uncontrolled self-motion.

6. The least-squares solution for redundancy resolution at the acceleration level was modified by adding a velocity-dependent term to the cost function. This modification successfully controlled the self-motion of the manipulator.

7. It was demonstrated by simulation that the force tracking performance of the methods based solely on inverse dynamics degrades in the presence of uncertainty in the manipulator's dynamic parameters and unmodelled dynamics. This is especially true for a manipulator equipped with harmonic drive transmissions, which introduce a high level of joint flexibility and frictional effects (as in the case of REDIESTRO).

8. The AHIC scheme has been modified by incorporating an "error reference controller". This modification successfully copes with model uncertainties in the model-based part of the controller, so that even friction compensation is not required.

In the next chapter, we illustrate further the capabilities of the AHIC scheme by showing expertimental results obtained using the REDIESTRO manipulator.

6 Experimental Results for Contact Force and Compliant Motion Control

6.1 Introduction

In this chapter, we describe the hardware experiments performed to evaluate the performance of the proposed AHIC scheme for compliant motion and force control of REDIESTRO. Considering the complexity and the large amount of calculations involved in force and compliant motion control of a 7-DOF redundant manipulator, the implementation of the real-time controller, from both hardware and software points of view, by itself represents a challenge. It should be noted that there are very few cases in the literature that experimental results for force and compliant motion control of a 7-DOF manipulator have been reported. In [67], a set of experiments on contact force control carried out on a 7-DOF Robotics Research Corporation (RRC) model K1207 arm at the Jet Propulsion Laboratory is reported. It should be noted that the RRC arm is one the most advanced manipulators from both mechanical design and controller viewpoints. On the other hand, implementation of the AHIC scheme for REDIESTRO introduces additional challenges:

- The REDIESTRO arm is equipped with harmonic drive transmissions which introduce a high level of joint flexibility. This makes accurate control of contact force more difficult.

- A friction model and its parameters cannot be estimated accurately in many practical applications. The friction model that is generally used models load independent Coulomb and viscous friction. This model is especially inadequate for a robot with harmonic drive transmissions which have high friction - experimental results show that in some configurations, the friction torques reach up to 30% of the applied torques. Also, experimental studies [88] have shown that frictional torques in harmonic drives are very nonlinear and load dependent. This represents a challenge for a model-based controller.

- Performing tasks such as "peg-in-the hole" requires very accurate positioning. This needs a very well-calibrated arm. In [15], Colombina *et al.* described the development of an impedance controller at the External Servicing Test-bed which is a ground test-bed currently installed at the European Space Agency Research Center. The performance of the impedance controller was demonstrated for a replacement of an Orbital Replacement Unit (ORU). They reported that only misplacement of 5 mm in position and 0.5 degrees in orientation are compensated for in an ORU exchange task. Considering the fact that REDIESTRO has not been accurately calibrated, the successful operation of the peg-in-the-hole strawman task by REDIESTRO demonstrates a high level of robustness of the proposed scheme.

The goal of this chapter is to demonstrate the feasibility and to evaluate the experimental performance of the control scheme described in the preceding chapters. Before presenting the experimental results, a detailed analysis is given to provide guidelines in the selection of the desired impedances. A heuristic approach is described which enables the user to systematically select the impedance parameters based on stability and tracking requirements.

At this stage different scenarios have been considered and two strawman tasks - surface cleaning and peg-in-the-hole - have been selected. The selection is based on the ability to evaluate force and position tracking and also robustness with respect to knowledge of the environment and kinematic errors. Finally, experimental results for these strawman tasks are presented. The hardware configuration (see Figure 6.1) used for the experimental work was developed to meet the requirements for force and compliant motion control.

6.2   Preparation and Conduct of the Experiments

6.2.1   Selection of Desired Impedances

The desired equation of motion in a position (impedance)-controlled direction is given by:

$$m^d \ddot{e} + b^d \dot{e} + k^d e = -f_e \qquad (6.2.1)$$

where $e = x - x^d$. The desired equation of motion in a force-controlled direction is given by:

$$m^d\ddot{x} + b^d\dot{x} = f^d - f_e \qquad (6.2.2)$$

The environment is modeled as a linear spring. Therefore, the interaction force in (6.2.2) can be replaced by $f_e = k_e x$, which results in

$$m^d\ddot{x} + b^d\dot{x} + k_e x = f^d \qquad (6.2.3)$$

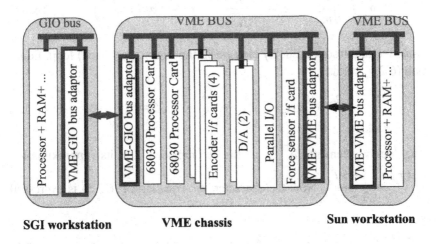

Figure 6.1 Hardware configuration (for force control experiments)

Comparing the desired equation of motion in a position (impedance) controlled direction (6.2.1) with that of a force-controlled direction (6.2.3), we note that the same guidelines for selection of impedance gains which ensure both stability and tracking performance can be used. The main difference is that in an impedance-controlled direction, the stiffness is an adjustable control parameter which can be specified while in a force-control direction, the stiffness is an environmental parameter which is not selectable. A complete stability analysis study and guidelines for selecting the set of impedance parameters to ensure stability of motion taking into account delays in the force and position sensor loops and also stiffness of contact are given in this section.

6.2.1.1 Stability Analysis

As mentioned above, the same guidelines can be followed for both impedance- and force-controlled directions. Therefore, we consider the following generic system:

$$m\ddot{x} + b\dot{x} + kx = f \tag{6.2.4}$$

Equation (6.2.4) can be expressed (using Laplace transforms) as

$$s^2 X(s) + 2\xi\omega_n s X(s) + \omega_n^2 X(s) = F(s) \tag{6.2.5}$$

where

$$\omega_n = \sqrt{\frac{k}{m}}, \qquad \xi = \frac{b}{2\sqrt{km}}, \qquad F = Laplace\left(\frac{f}{m}\right) \tag{6.2.6}$$

Now, let us introduce a delay element in the sensor (feedback) loop. Equation (6.2.5) yields

$$s^2 X + 2\xi\omega_n e^{-2T_s s} s x + \omega_n^2 e^{-2T_s s} X = F \tag{6.2.7}$$

The delay element $e^{-2T_s s}$ can be replaced by its approximation $e^{-2T_s s} = \dfrac{1 - sT_s}{1 + sT_s}$. Now the characteristic equation of (6.2.5) is expressed by:

$$T_s s^3 + (1 - 2\xi\omega_n T_s)s^2 + (2\xi\omega_n - \omega_n^2 T_s)s + \omega_n^2 = 0 \tag{6.2.8}$$

According to the Routh stability criterion, the system expressed by (6.2.7) is stable (all roots of (6.2.8) are in the left-half of the complex plane) if and only if all coefficients in the first column of the Routh table have the same sign. This leads to

$$\omega_n < \frac{2\xi}{T_s} \qquad and \qquad \omega_n < \frac{1}{2\xi T_s} \tag{6.2.9}$$

6.2.1.2   Impedance-controlled Axis

The desired equation of motion is given by (6.2.1). In this case, the desired mass, damping, and stiffness should be specified. The following steps are required:

- Based on the sampling and sensor delays, select ξ and ω_n such that the stability condition (according to Figure 6.2) is satisfied.

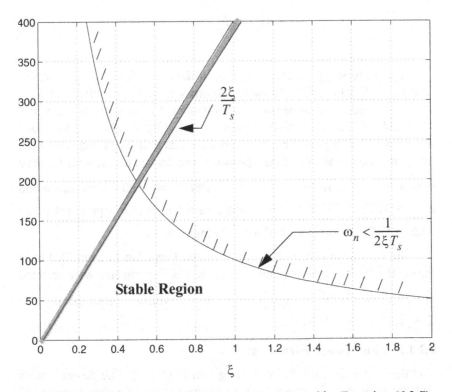

Figure 6.2 Stability region of the system represented by Equation (6.2.7) with T_s = 0.005 seconds.

- Select the desired stiffness according to the acceptable steady-state error:

$$e_{ss} = \frac{-f_e}{k^d} \qquad (6.2.10)$$

where f_e is the disturbance force in a position-controlled direction such as the friction force on the surface for a surface cleaning scenario.

- Calculate the desired inertia and damping using:

$$m^d = \frac{k^d}{\omega_n^2} \qquad (6.2.11)$$

$$b^d = \frac{2\xi k^d}{\omega_n} \qquad\qquad (6.2.12)$$

In order to study the step response of the controller in an impedance-controlled direction, the following experiment was conducted. All axes were specified to be impedance-controlled for the segment between $t =$ 110s and $t =$115s. The desired position trajectory is specified such that there is a difference of 13 cm between the initial desired position along the z axis and the initial tool frame z position. The desired impedances for the z axis are specified by: $m^d = 112, b^d = 700, k^d = 1100$ which correspond to $\zeta = 1, T_n = 2s$. Figure 6.3 compares the hardware experiment result with that of the ideal system of mass-spring-dashpot.

The desired impedances for the position (impedance)-controlled axes during the surface cleaning and the peg in hole experiments were selected as $m^d = 257, b^d = 1100, k^d = 1100$ which correspond to $\zeta = 1.03, T_n = 3.03s$.

6.2.1.3   Force-controlled Axis:

The desired equation of motion is given in (6.2.2). The desired mass and damping should be specified. In contrast to an impedance-controlled axis where the stiffness is an adjustable control parameter, in this case the stiffness k_e is the overall stiffness of contact. The contact stiffness is affected by the following factors:

- Tool stiffness: the eraser pad in the case of surface cleaning and the plexi-glass peg in the case of peg in the hole.

- Environment stiffness: the white-board table and its support in the case of surface cleaning and the plexi-glass hole in the case of peg in the hole.

- Transmission (joint) flexibility: the flexibility of harmonic drives.

- Structural (link) flexibility

Therefore, in order to assign ξ and ω_n for the force-controlled axis, one should know the overall stiffness of contact. Although difficult to determine, the stiffness of the tool and environment can be identified by off-line experiments; joint and link flexibilities are even more difficult to

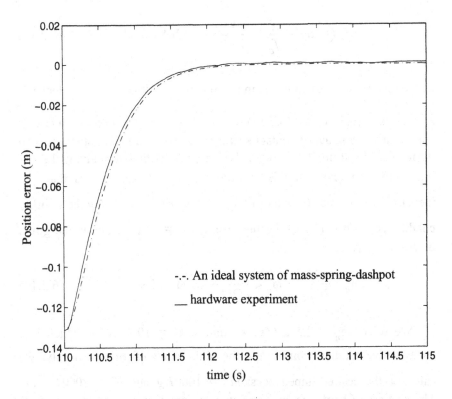

Figure 6.3 Position step response in an impedance-controlled direction (13 cm initial position error).

identify and characterize. Note that the force tracking steady-state error in (6.2.2) is not affected by the stiffness k_e as long as the system remains stable. However, the transient response varies with k_e. In conducting the experiments, a heuristic approach has been used which allows us to achieve the desired steady-state and transient performance without an elaborate procedure to identify and characterize the overall stiffness of contact.

Based on the estimate of the delay in the force sensor loop, we select ξ and ω_n such that the stability condition according to Figure 6.2 is satisfied. The major delay in this case is due to the low-pass force sensor filter with cutoff frequency f_c equal to 7.81 Hz. The filter delay is approximately given by:

$$Delay = \frac{1}{f_c} = \frac{1}{7.81} = 0.128S \qquad (6.2.13)$$

Based on a very conservative estimate of contact stiffness k_e, we select m^d and b^d as in (6.2.11) and (6.2.12) respectively. Note that in order to have a conservative estimate of contact stiffness, we select a higher stiffness than what would be normally expected. This can be justified by studying the stability criterion given in this section. Equation (6.2.6) shows that ω_n increases with increasing values of k_e with the risk of violating the stability conditions given in (6.2.9). In this case, for $\zeta = 1$, the stability margin is determined by:

$$\omega_n < \frac{1}{2\zeta T_s} \Rightarrow \omega_n < \frac{1}{2Delay} \Rightarrow \omega_n < 3.9 \qquad (6.2.14)$$

We select $\omega_n = 2.5rad/s$. Assuming $k_e = 10000N/m$ which is a conservative estimate considering the high values of joint flexibility, we calculate the desired impedances: $m^d = 1600Kg$ and $b^d = 8000Ns/m$. The first set of hardware experiments were conducted using these impedances.

The test scenario (Figure 6.4) consists of the first three segments of the surface cleaning strawman task (see Section 6.3.1). In the first segment, the eraser pad is positioned above the white-board table (all axes under position control) in 5 seconds. In the second segment, the eraser approaches the surface along the z axis under force control, while keeping the position along the x and y axes fixed. The desired force along the z axis is 20N. In the last segment, the eraser is commanded to move along the y axis on the surface with a desired 20N force.

Figure 6.5 shows the plot of the interaction forces. The response time of the system (~10s) is greater than the expected value (2.53s based on $k_e = 10000\frac{N}{m}$), which shows that the actual stiffness of the contact is much less than the estimated value.

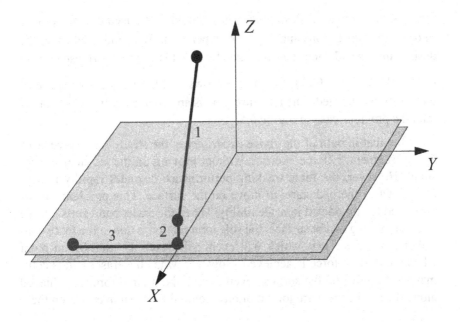

Figure 6.4 Test scenario for selecting the desired impedance in the force-controlled direction for the first strawman task

The heuristic approach of selecting the desired impedances is based on studying the actual response in hardware experiments. As an example, let us calculate a set of impedances which results in a response time that is twice as fast (5 seconds) compared to the 10 seconds in Figure 6.5. Assuming a fixed contact stiffness k_e, Equation (6.2.6) results in

$$\frac{\omega_{n_1}}{\omega_{n_2}} = \sqrt{\frac{m_2}{m_1}} = \frac{T_{n_2}}{T_{n_1}} \tag{6.2.15}$$

Equation (6.2.15) suggests that in order to reduce T_n by a factor of 2, the desired mass should be reduced by a factor of 4. Equation (6.2.6) also results in

$$\frac{\xi_1}{\xi_2} = \sqrt{\frac{m_2}{m_1}} \cdot \frac{b_1}{b_2} \tag{6.2.16}$$

which suggests that the desired damping should be reduced by a factor of 2 in order to keep ξ constant. The next experiment was conducted using the desired inertia and damping calculated by (6.2.15) and (6.2.16) respectively (m^d = 400, b^d = 4000). Figure 6.6 shows that the transient response of the system is changed. The response time is approximately two times faster than the previous case in Figure 6.5.

Note that in both of the above experiments, the steady-state force error during segment 2 (force exertion without moving on the surface) is very small. However, the force tracking performance degrades rapidly in segment 3 when the pad starts to move on the surface. This problem can be attributed to unmodeled joint flexibility. When the eraser pad exerts a force without moving on the surface, the sole joint motion is due to the force controller along the z axis which will eventually reach an equilibrium point when the desired force is achieved. However, when the eraser pad is commanded to move on the surface, even though the desired force is achieved along the z axis, there are joint motions required for the movement on the

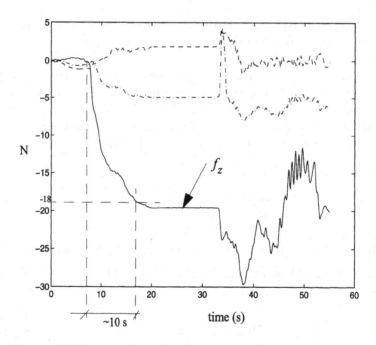

Figure 6.5 Force tracking for the test case shown in Figure 6.4 with
$$m^d = 1600, b^d = 8000.$$

surface. Without the unmodeled dynamics due to joint flexibility, the motions along the position-controlled directions and the force-controlled directions are decoupled. Therefore, the horizontal movement on the surface should not affect the force tracking along the z axis. However, any joint oscillation due to unmodeled joint flexibility acts as a coupling between the position-controlled directions and the force-controlled directions which causes performance degradation in force tracking (see Figure 6.5).

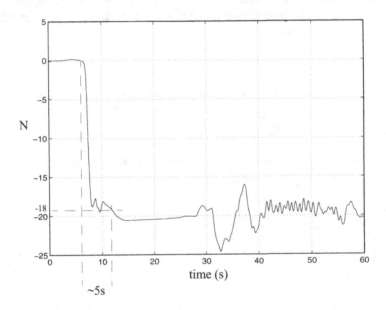

Figure 6.6 Force tracking for the test case shown in Figure 6.4 with
$$m^d = 400, b^d = 4000 \quad .$$

The force controller in (6.2.2) can be seen as a second-order filter (see Figure 6.7) with a corner frequency of $\sqrt{k_e/m^d}$. Now, in order for the force-controller in (6.2.2) to reject these disturbances (the forces due to joint oscillations), the cutoff frequency of this filter should be selected much greater than the frequency of the disturbances (in this case oscillations caused by joint flexibility). In order to increase the cutoff frequency, one should reduce the desired inertia (m^d) to as small a value as possible while maintaining system stability. The values of the desired inertia and damping were selected experimentally as $\quad m^d = 5.7, b^d = 477 \quad .$

6.2.2   Selection of PD Gains

In the modified AHIC scheme (see Figure 5.21), a PD controller was implemented to ensure that the reference error (error between the target trajectory generated by the AHIC controller and tool frame trajectory) converges to zero. Therefore, in order for the robot to act as closely as possible to the ideal impedance system specified by (6.2.1) and (6.2.2), the PD gains need to be selected as high as possible. Different experiments were conducted to find the best values for the PD gains. The maximum values that do not excite the unmodeled dynamics were obtained experimentally as $k_p = 400, k_v = 40$.

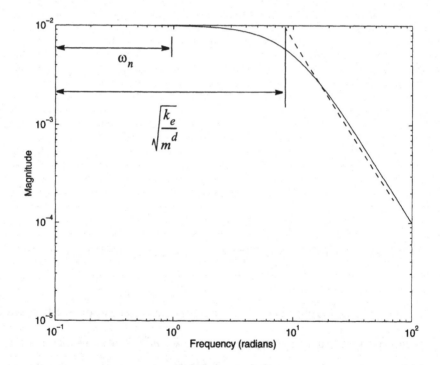

Figure 6.7 Plot of magnitude versus frequency for the second-order filter in Equation (6.2.3).

6.2.3 Selection of the Force Filter

The force sensor data usually contain a high level of noise which needs to be filtered out by implementation of a low-pass filter. The selection of the filter is a trade-off between noise rejection and stability requirements as the low-pass filter introduces a delay in the sensor loop which can cause instability. The JR3 force-sensor interface card provides a cascade of low-pass filters. Each succeeding filter has a cutoff frequency that is 1/4 of that of the preceding filter. For the JR3 sensor with a sample rate of 8 kHz, the cutoff frequency of the first filter is 500Hz. The subsequent filters cutoff at 125 Hz, 31.25 Hz, 7.81 Hz and so on. The optimal filter has been selected experimentally. Figure 6.8 shows the force measurements with different filters for the test scenario of Figure 6.4. As one may notice, the filter with

7.81 Hz cutoff frequency gives the best tracking (f_z^d = $15N$) with maximum noise reduction.

6.2.4 Effect of Kinematic Errors (Robustness Issue)

The AHIC scheme may suffer from two major sources of kinematic errors:

- **The kinematic parameters of the arm**: In the absence of an accurate kinematic calibration, the forward kinematics based on kinematic parameters can introduce errors in the calculated Cartesian feedback.

- **The robot's environment**: The kinematic description of the robot's environment, such as position and orientation of the constraint frame, introduces kinematic errors when the Cartesian feedback is transformed into the constraint frame.

Different solutions may be envisaged

1– Kinematic calibration of the arm.

2– Kinematic calibration of the environment.

3– Use of a real-time vision system.

4– Mechanical design of the tool attachment.

5– Exploitation of the capabilities of the AHIC scheme.

The Cartesian feedback (linear and angular position and rates) is calculated based on the joint angles and the forward kinematics of the manipulator. Therefore, accurate kinematic calibration can improve the performance. Kinematic calibration of the environment (robot's base coordinates and the

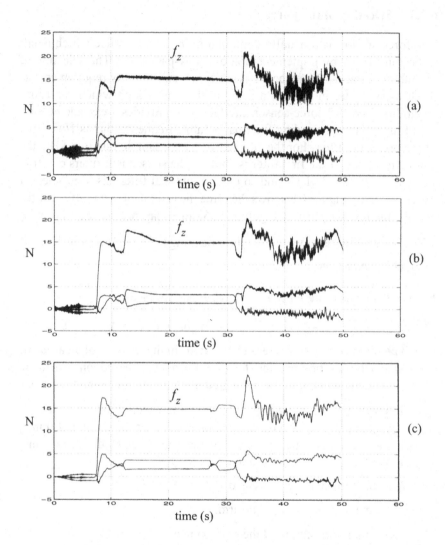

Figure 6.8 Effect of the force filter cutoff frequency;
a) 125 Hz, b) 31.25 Hz, and c) 7.81 Hz

constraint surfaces) will also improve the performance. An alternative to
kinematic calibration is to use a real-time vision system which gives the
appropriate feedback expressed in a desired frame. Mechanical design of
the tool attachment can also play a important role in performance improve-
ment. For instance, using a universal joint in the surface cleaning demonstration improves the performance by rejecting interaction torques which

would otherwise be present due to surface orientation errors (or errors in arm calibration). The AHIC scheme itself can act as a tool to deal with kinematic errors at two levels:

- The impedance controller in the position-controlled directions can gracefully handle any coupling forces (disturbances) due to kinematic errors.

- The force/torque controller uses only force sensor feedback (the linear and angular position and rate feedback does not appear in the force/torque-controlled directions). This force sensor data provides error-free information about the kinematics of the environment and constraint surfaces.

In conducting the strawman tasks, we have relied solely on solutions 4 and 5 (mechanical design of the tool attachment and on exploiting the AHIC scheme). This emphasizes the performance of the controller and its robustness with respect to kinematic errors. As an example, the design of the peg and holes (cone-shaped peg heads and chamfered type opening at the top of the holes) can accommodate certain position errors due to imprecise kinematic parameters of the arm and its environment. Before presenting the numerical results of the strawman tasks, let us study the performance of the AHIC scheme in identifying the correct kinematics of the environment using force sensor feedback without relying on knowledge of the kinematics of the arm and its environment.

In the following experiment, a plexi-glass rectangular plate was rigidly attached to the last link such that the plate's normal is parallel to the tool frame's x axis. The task consists of two segments: in the first segment the manipulator is commanded to go in five seconds to a position above the surface with the tool frame's x axis perpendicular to the surface ($s = diag[1, 1, 1, 1, 1, 1]$); in the second segment, the position along the x and y axes (see Figure 6.4) is kept constant while the desired force along the z axis is specified (20 N). All three rotational axes are specified to be torque-controlled to deal with any misorientation of the surface ($s = diag[1, 1, 0, 0, 0, 0]$). Note that the position and orientation of the task frame (in this case, the surface frame) are provided by the user. Also, in the following sections, the impedance- or force-controlled directions are specified by $s = diag[p_x, p_y, p_z, r_x, r_y, r_z]$ where a 0 entry indicates a force/torque-controlled direction, and a 1 entry indicates an impedance-controlled direction.

In order to study the robustness of the algorithm, we introduce $\sim 5°$ (along each rotational axis) of orientation error on the surface (see Figure 6.9). Figure 6.10 shows that there is considerable torque at the initial stage of contact with the surface – this is due to the orientation mismatch. This initial torque reduces very rapidly because the controller tries to regulate the torques to zero. Hence, the plate detects the correct orientation of the surface. We can also see the performance of the force controller in regulating the normal force to the surface to 20N. Note that this experiment is similar to that of inserting a peg into a hole when the peg and the hole axes are not completely aligned. Therefore, the desired mass and damping for the three rotational axes, $m^d = 0.056, b^d = 48$, can also be used for the second strawman task (peg in the hole).

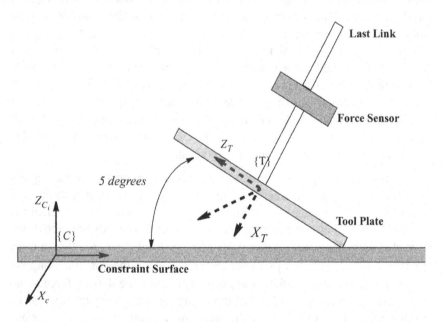

Figure 6.9 Hardware experiment to illustrate the capability of the AHIC scheme in identifying the correct kinematics of the environment using a force sensor

6.3   Numerical Results for Strawman Tasks

The hardware and software environment used for the experimental work also allows for visualization of the motion of the arm on an SGI workstation running the MRS graphical software environment (see Figure 6.11). The joint angles as well as the force/torque sensor data are transferred in real time to the SGI workstation

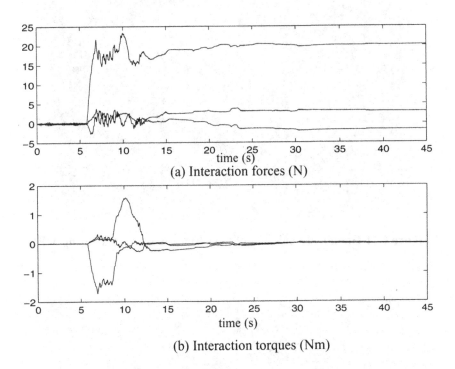

(a) Interaction forces (N)

(b) Interaction torques (Nm)

Figure 6.10 Experimental results: Exerting 20N on the surface with a rigid plate, with 5 degrees (along each rotational axis) of orientation error on the surface.

6.3.1   Strawman Task I (Surface Cleaning)

Figure 6.12 shows a perspective view of the setup used for the hardware demonstration of the strawman task. The five segments of the task are shown in Figure 6.13. Table 6-1 summarizes the control parameters used for this task, where "y" denotes information that is not needed. The desired masses for the position and the force-controlled directions are $M_p = 257$ and $M_f = 5.7$ respectively. The values of the desired damping in the position and the force-controlled directions are $B_p = 1100$ and $B_f = 477$; and the desired stiffness in the position-controlled direc-

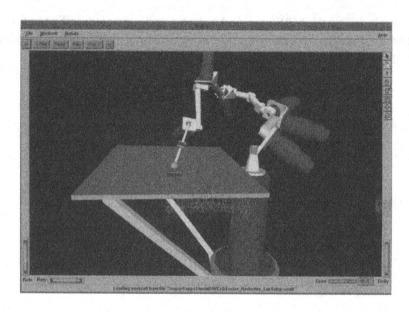

Figure 6.11 Graphical rendering of the surface-cleaning task using MRS

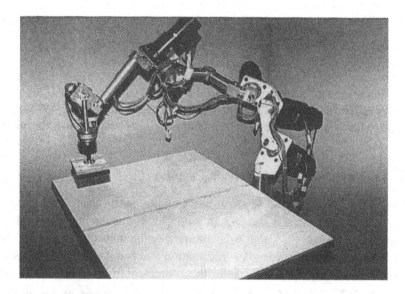

Figure 6.12 Perspective view of the hardware setup used in the
demonstration of Strawman Task I (surface cleaning)

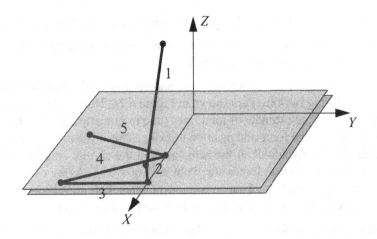

Figure 6.13 Different segments of Strawman Task I (surface cleaning)

Table 6-1 Control parameters used for Strawman Task I

Seg	T (s)	Selection vector	Desired Inertia	Desired damping	Desired stiffness	Desired Force/ torques
1	5	1 1 1 1 1 1	$M_p M_p M_p$ $M_p M_p M_p$	$B_p B_p B_p$ $B_p B_p B_p$	$K_p K_p K_p$ $K_p K_p K_p$	y^a y y y y y
2	25	1 1 0 1 1 1	$M_p M_p M_f$ $M_p M_p M_p$	$B_p B_p B_f$ $B_p B_p B_p$	$K_p K_p$ y $K_p K_p K_p$	y y -20 y y y
3	50	1 1 0 1 1 1	$M_p M_p M_f$ $M_p M_p M_p$	$B_p B_p B_f$ $B_p B_p B_p$	$K_p K_p$ y $K_p K_p K_p$	y y -20 y y y
4	75	1 1 0 1 1 1	$M_p M_p M_f$ $M_p M_p M_p$	$B_p B_p B_f$ $B_p B_p B_p$	$K_p K_p$ y $K_p K_p K_p$	y y -20 y y y
5	100	1 1 0 1 1 1	$M_p M_p M_f$ $M_p M_p M_p$	$B_p B_p B_f$ $B_p B_p B_p$	$K_p K_p$ y $K_p K_p K_p$	y y -20 y y y

a. "y" denotes information that is not needed.

tion is $K_p = 1100$. The PD gains are selected as $k_p = 400, k_v = 40$. Also note that in this experiment, no joint-friction compensation in the inverse dynamics module is performed.

As an example, the joint angle, rate and commanded torque for joint 2 (gathered during run-time) are shown in Figure 6.14. Note that the presence of the noise on the estimate of joint rates (due to numerical differentiation) has not affected the force and position tracking. This noise is effectively filtered out by the dynamics of the actuators and the current amplifiers. The results of the interaction forces are given in Figure 6.15. As we can see, the force tracking in the 5 to 25 seconds segment when there is no motion in the x and y directions is almost perfect (0.04 N steady-state error).

It was noted in Section 6.2.1.3 that when the pad moves on the surface, the force tracking can degrade drastically because of unmodeled flexibility in the joints. However, by appropriate selection of the controller's cutoff frequency in a force-controlled direction (see Figure 6.8), one can achieve an acceptable level of force tracking. For the segments beyond 25 seconds, there is a low amplitude (approximately 1 N) oscillation with a frequency around 1 Hz due to unmodeled joint flexibility. However, the mean value and standard deviation ($f_{mean} = -19.60N, f_{std} = 0.6$) for the time interval between 15 to 80 seconds show the capability of the force-controller in regulation of the interaction forces even in the presence of unmodeled dynamics (joint frictions and flexibilities).

There is also considerable friction on the surface: approximately 5N in the y direction and 2N in the x direction. However, the impedance controller is not only stable in these directions, but is also successful in achieving acceptable tracking with 1 cm steady-state error in the y direction and 0.5 cm in the x direction (see Figure 6.16). These errors can be further reduced by assigning a larger k^d in the impedance-controlled directions.

6.3.2   Strawman Task II (Peg In The Hole)

Figure 6.17 shows a perspective view of the setup used for the hardware demonstration of the task. The complete task consists of accomplishing the insertion of a peg into, and its removal from, two different holes. In Section 6.2.4, we described the effects of kinematic errors. It was noted that kinematic errors can play a vital role in performing tasks such as the peg-in-

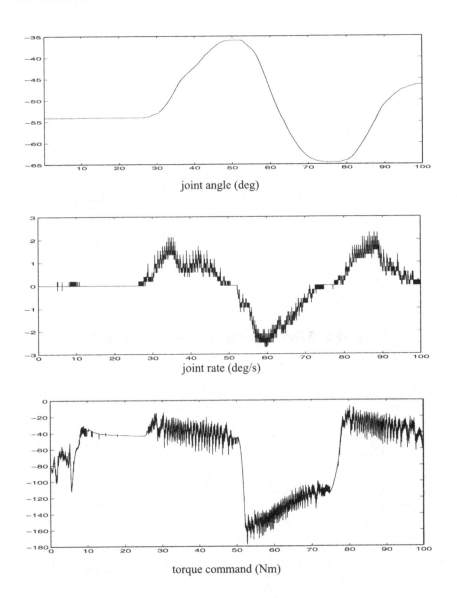

Figure 6.14 Strawman Task I: Captured data for joint 2

the-hole operation. In most cases kinematic errors result in conflicts between the position and force-controlled directions which can cause oscillations and instability. Different solutions have been suggested. However, in performing this strawman task, no calibration of the arm kinematics and

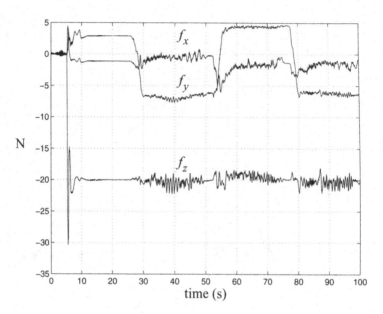

Figure 6.15 Force data captured for Strawman Task I

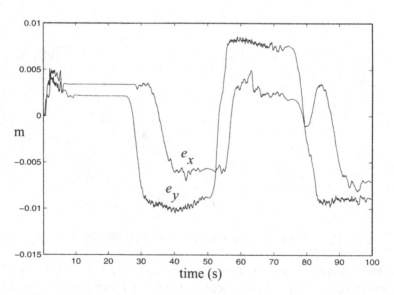

Figure 6.16 Position errors for the surface-cleaning hardware demonstration

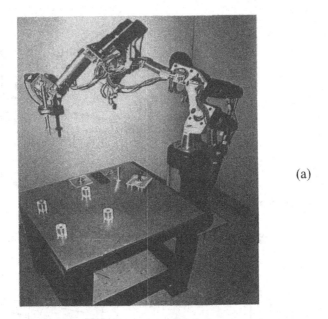

(a)

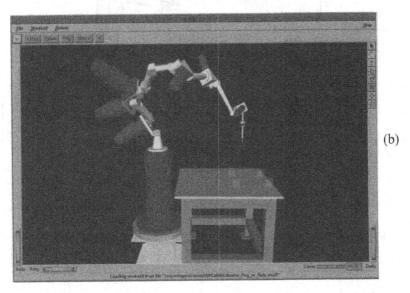

(b)

Figure 6.17 Perspective view of the hardware setup used in demonstration of Strawman Task II; a) The REDIESTRO arm, b) MRS simulation

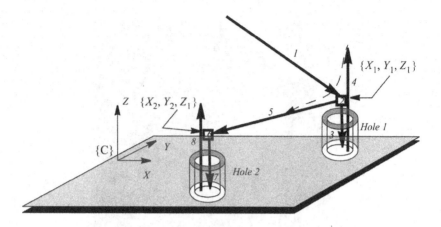

Figure 6.18 Different segments of Strawman Task II

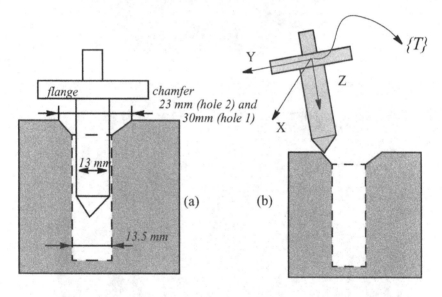

Figure 6.19 Strawman Task II: a) Kinematic tolerances, b) Correct
initiation of the insertion

the hole setup was performed. Instead, we have relied on the kinematic
design of the peg and the holes, and also on the capability of the AHIC
scheme to deal with kinematic errors. The dimensions of the peg and the
holes are given in Figure 6.19a. The final tolerance between the peg and the

holes is 0.25 mm. However, the structural designs of the head of the peg and the top of the holes facilitate the correct initiation of the insertion process in the presence of small positioning errors. With this design, only an accuracy/repeatability of 15mm for hole 1 and 11.5 mm for hole 2 is required to initialize the insertion (see Figure 6.19b). From this point onward, the AHIC scheme is responsible for accomplishing the insertion in the presence of kinematic errors.

The strawman task consists of 8 segments. Table 6-2 summarizes the desired position, orientation, and force for the task. The control parameters are given in Table 6-3. The descriptions of the different segments of the strawman task are as follows:

Table 6-2 Desired positions, orientations, and forces for Strawman Task II

Seg	T (s)	Selection vector	Desired Position (m)	Desired Orientation	Desired Force(N)	Des. Torque (Nm)
			(X,Y,Z)	(Rx,Ry,Rz)	(Fx,Fy,Fz)	(Tx,Ty,Tz)
1	3	1 1 1 1 1 1	$(x1^a,y1,z1)$	$(pi/2,pi/2,0)$	(NA,NA,NA)	(NA,NA,NA)
2	10	0 0 1 1 1 1	$(NA^b,NA,z1)$	$(pi/2,pi/2,0)$	(0,0,NA)	(NA,NA,NA)
3	65	0 0 0 0 0 0	(NA,NA,NA)	(NA,NA,NA)	(0,0,-8)	(0,0,0)
4	110	0 0 0 0 0 0	(NA,NA,NA)	(NA,NA,NA)	(0,0,10)	(0,0,0)
5	115	1 1 1 1 1 1	(x2,y2,z1)	$(pi/2,pi/2,0)$	(NA,NA,NA)	(NA,NA,NA)
6	125	0 0 1 1 1 1	(NA,NA,,z1)	$(pi/2,pi/2,0)$	(0,0,NA)	(NA,NA,NA)
7	190	0 0 0 0 0 0	(NA,NA,NA)	(NA,NA,NA)	(0,0,-8)	(0,0,0)
8	225	0 0 0 0 0 0	(NA,NA,NA)	(NA,NA,NA)	(0,0,10)	(0,0,0)

a. See Figure 6.18
b. NA =Not Applicable

Table 6-3 Control parameters used in Strawman Task II

Seg	T (s)	Selection vector	Desired Inertia	Desired damping	Desired stiffness	Desired Force/torques
1	3	1 1 1 1 1 1	$m_p{}^a\, m_p\, m_p$ $m_p\, m_p\, m_p$	$b_p\, b_p\, b_p\, b_p$ $b_p\, b_p$	$k_p\, k_p\, k_p\, k_p$ $k_p\, k_p$	y^b y y y y y
2	10	0 0 1 1 1 1	$m_{f1}\, m_{f1}\, m_p$ $m_p\, m_p\, m_p$	$b_{f1}\, b_{f1}\, b_p\, b_p$ $b_p\, b_p$	$y\, y\, k_p\, k_p\, k_p$ k_p	0 0 y y y y
3	65	0 0 0 0 0 0	$m_{f1}\, m_{f1}\, m_{f2}$ $m_{f3}\, m_{f3}\, m_{f3}$	$b_{f1}\, b_{f1}\, b_{f2}\, b_{f3}$ $b_{f3}\, b_{f3}$	y y y y y y	0 0 -8 0 0 0
4	110	0 0 0 0 0 0	$m_{f1}\, m_{f1}\, m_{f2}$ $m_{f3}\, m_{f3}\, m_{f3}$	$b_{f1}\, b_{f1}\, b_{f2}\, b_{f3}$ $b_{f3}\, b_{f3}$	y y y y y y	0 0 10 0 0 0
5	115	1 1 1 1 1 1	$m_{p1}\, m_{p1}\, m_{p1}$ $m_{p1}\, m_{p1}\, m_{p1}$	$b_{p1}\, b_{p1}\, b_{p1}$ $b_{p1}\, b_{p1}\, b_{p1}$	$k_p\, k_p\, k_p\, k_p$ $k_p\, k_p$	y y y y y y
6	125	0 0 1 1 1 1	$m_{f1}\, m_{f1}\, m_p$ $m_p\, m_p\, m_p$	$b_{f1}\, b_{f1}\, b_p\, b_p$ $b_p\, b_p$	$y\, y\, k_p\, k_p\, k_p$ k_p	0 0 y y y y
7	190	0 0 0 0 0 0	$m_{f1}\, m_{f1}\, m_{f2}$ $m_{f3}\, m_{f3}\, m_{f3}$	$b_{f1}\, b_{f1}\, b_{f2}\, b_{f2}$ $b_{f3}\, b_{f3}$	y y y y y y	0 0 -8 0 0 0
8	225	0 0 0 0 0 0	$m_{f1}\, m_{f1}\, m_{f2}$ $m_{f3}\, m_{f3}\, m_{f3}$	$b_{f1}\, b_{f1}\, b_{f2}\, b_{f3}$ $b_{f3}\, b_{f3}$	y y y y y y	0 0 10 0 0 0

a. see Table 6-4 for numerical values
b. y = not needed

Table 6-4 Numerical values of the desired impedances (Strawman Task II)

Position-controlled axis			Force-controlled axis	
desired mass (kg)	desired damping Nsec/m	desired stiffness N/m	desired mass (kg)	desired stiffness N/m
(m_p, m_{p1})	(b_p, b_{p1})	k_p	(m_{f1}, m_{f2}, m_{f3})	(b_{f1}, b_{f2}, b_{f3})
(257, 112)	(1100, 700)	1100	(22.8, 253, 0.056)	(955, 3180, 48)

Segment 1: In this segment the tool frame {T} is commanded to go to position $\{x_1, y_1, z_1\}$, where x_1 and y_1 are the x and y coordinates of the center of hole 1 as seen in {C}; z_1 is specified to ensure that the peg's head is above the hole's upper surface. The desired orientation is specified such that the x axis of {T} is aligned with the axis of hole 1 (see Figure 6.13). Note that because of kinematic errors, the position and orientation of the tool frame are different from their desired values.

Segment 2: It was noted that due to the presence of different sources of kinematic errors, the position and orientation of the tool frame would be different from the desired values at the end of segment 1. In segment 2, the possibility of manually correcting the tool frame position in the x and y directions is given to the operator. This is only required if the kinematic errors are such that the insertion cannot be initiated correctly (see Figure 6.19b). In this case, the operator can drag the peg to a position from where the insertion can be initialized. This is done by keeping the orientation and the tool frame's height (along the z axis) constant while the x and y axes are under force control with zero desired force. In the hardware experiment for Strawman Task II, no manual correction was needed.

Segment 3 (insertion): In this segment all axes are under force/torque control. In this way the force/torque sensor information is used to accurately align the axes of the peg and the hole. Only a negative desired force along the z axis is specified The desired forces/torques for the other axes are zero. Note that no logic branching is required to detect the end of the insertion. The motion is stopped upon completion of the insertion, i.e., on achieving the desired interaction force between the peg's flange and the top surface of the hole.

Segment 4 (removal): This segment is similar to segment 3, with the difference that the desired force is in the positive z direction to accomplish the removal.

Segment 5: This is the transmission segment to locate the peg on top of hole 2. Note that in segment 4, the z axis was under force control attempting to achieve a positive force. Because there is no constraint on the tool frame that allows the desired force to be achieved, the tool frame continues to move along the positive z direction with a bounded terminal velocity according to a time-controlled schedule. By starting segment 5, all axes come under position control so as to position the peg on top of the second hole. As noted in Section 5.2.1 , the task planner module uses a pre-specified task file to calculate the coefficients of the desired trajectory for different segments before starting the task. Therefore, the initial position of the tool frame, i.e., the final position at the end of segment 4, is not known ahead of time. In this experiment, we used the desired final position for segment 1 as the initial position of segment 5. Therefore, there is an initial position error when segment 5 starts. However, as mentioned in Section 6.2.1.2, this does not cause any difficulties since the impedance controller smoothly "attracts" the tool frame to the desired trajectory (see dashed-line in Figure 6.18).

Segment 6: Similar to segment 2.

Segment 7: Similar to segment 3.

Segment 8: Similar to segment 4

Figure 6.20 and Figure 6.21 show the results of the hardware experiment for Strawman Task II. In order to get a better resolution, only the insertion and removal procedures for hole 1 are shown. The following phases can be observed in Figure 6.20:

Phase 1 (position correction): When the head of the peg touches the chamfer at the top of the hole, the interaction forces in the x and y direction modify the position mismatch (due to kinematic errors) and guide the head of the peg into the hole. This happens because the x and y coordinates of the $\{T\}$ frame are force-controlled. As one can see, the interaction forces between the head of the peg and the body of the chamfer are reduced as the center of the peg enters the hole. The plot of the y coordinate for this phase shows the position modification (approximately 5mm).

Phase 2 (orientation correction): As the peg is inserted further into the hole, there is considerable force/torque build up because of the misalignment of the peg and the hole. This reduces rapidly as the torque controller for all three rotational axes reacts to modify the alignment of the peg. The interaction forces and torques become smaller when correct alignment is achieved.

Phase 3 (completing the insertion): After the peg's flange touches the top surface of the hole, the force controller tries to regulate the force in the z direction to the desired value ($-8N$). At this point ($t = \sim50s$) there are minimum interaction torques around all three rotational axes and forces in the x and y directions. This shows correct positioning and alignment of the peg. Note that at this stage, no logic (mode) branching is required. The peg remains inserted until the removal phase starts.

Phase 4 (removal): In this phase, a positive desired force is specified which forces the removal process to start.

In order to test the peg-in-the hole operation in the case of a tight fitting scenario, a layer of aluminum foil was wrapped around the peg which prevented the peg from sliding freely (under its own weight) into the hole. Strawman Task II was successfully demonstrated for this case as well. The only parameter that needed to be modified was the desired force in the z direction which was increased from 8N to 15N. This was necessary to prevent the peg from jamming.

In order to test the robustness of the scheme with respect to the kinematic description of the environment, the above scenario was repeated while introducing $\sim 5°$ orientation error on the axes of the holes. Strawman Task II was successfully demonstrated for this case as well.

6.4 Conclusions

The goal of this chapter was to demonstrate the feasibility and to evaluate the performance of the proposed compliant motion and force control scheme via hardware demonstrations using REDIESTRO, a seven-dof experimental robot arm. Two strawman tasks – surface cleaning and peg-in-the-hole – were selected. The results for the surface cleaning strawman task indicate that when there is no motion in the x and y directions, the force tracking is almost perfect (0.04 N steady-state error). When the eraser is moving on the surface, it was observed that because of unmodeled flexibility in the joints, the force tracking tends to degrade. However, by appropriate selection of the controller's cutoff frequency in a force-controlled direction, we achieved an acceptable level of force tracking. The experiment shows that with 20N desired force, an interaction force with mean value -19.6N and standard deviation 0.6 is achieved. This demonstrates the capability of the force-controller in regulating interaction forces even in the presence of unmodeled dynamics (joint frictions and flexibilities).

Strawman Task II was also successfully demonstrated. Considering the tolerances used for the design of the peg and the holes (0.25 mm between the radius of the peg and the hole) and the fact that REDIESTRO has not been kinematically calibrated, the successful peg-in-the-hole demonstration illustrates the robustness of the proposed scheme with respect to poor knowledge of the environment.

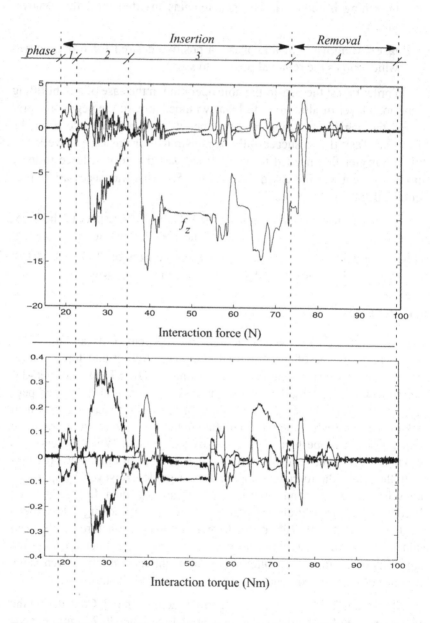

Figure 6.20 Strawman Task II: Hole 1 insertion/removal; interaction force/torques

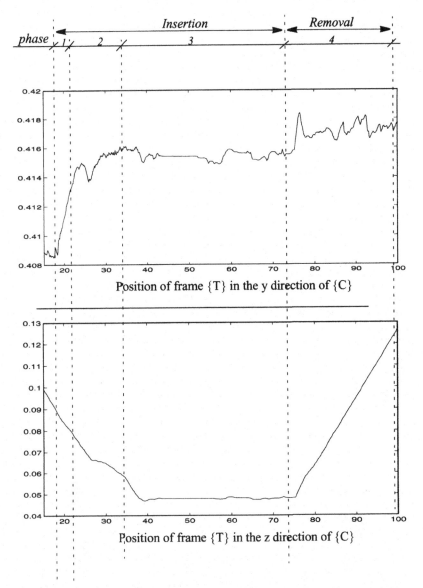

Figure 6.21 Strawman Task II: Hole 1 insertion/removal; position information

7 Concluding Remarks

As indicated in the Chapter 1, the objectives of this monograph were to present a unified framework for combining compliant motion control, redundancy resolution, and adaptive control in a single methodology and to demonstrate the the feasibility of the proposed scheme by computer simulations and experiments on REDIESTRO, a seven-dof experimental redundant manipulator. These objectives were achieved as follows:

The basic issues needed for analysis of kinematically redundant manipulators were presented in Chapter 2. Different redundancy resolution schemes were reviewed. Based on this review, *configuration control* at the acceleration level was found to be the most appropriate approach for force and compliant motion control of redundant manipulators. A formulation of the additional tasks to be used for redundancy resolution was presented. Joint limit avoidance, one of the most useful additional tasks to avoid mechanical joint limits, and self-collision avoidance, were studied in greater detail. The basic formulation of static and moving obstacle collision avoidance tasks in 2D workspace was presented.

The extension of redundancy resolution and obstacle avoidance scheme to the 3D workspace of REDIESTRO was addressed in Chapter 3. The obstacle avoidance algorithm was modified to consider 3-D objects. A novel primitives-based collision avoidance scheme was presented. This scheme is general and provides realism, efficiency of computation, and economy in preserving the amount of free space that would otherwise be wasted. Possible cases of collisions were also considered. In particular, cylinder-cylinder collision avoidance which represents a complex case for a collision detection scheme was formalized using the notions of dual vectors.

Before performing experimental work using REDIESTRO to evaluate the performance of the redundancy resolution and obstacle avoidance scheme, extensive simulations were performed using the kinematic model of REDIESTRO. The simulation results indicate that the least-squares approach for redundancy resolution is important for practical applications in order to cop with kinematic and artificial singularities. The latter may arise because of conflicts between the main and additional tasks. However, this introduction of singularity robustness results in tracking errors in regions away from singularities. It has been shown that by a proper selection (or time-varying formulation) of W_v, the weighting matrix in the singularity robustness task, the effect of this term on the tracking performance can be minimized. It was also shown that the formulation of the additional task as an inequality constraint, may result in considerable discontinuity in joint velocities which causes a large pulse in joint accelerations. In a practical implementation, the maximum acceleration of each joint would be limited, and this commanded joint acceleration would result in saturation of the actuators. A time-varying formulation of the weighting matrix, W_c, was proposed which successfully overcame this problem.

Three scenarios encompassing most of the developed redundancy resolution and obstacle avoidance system features were successfully demonstrated on real hardware, i.e. the REDIESTRO manipulator. These scenarios verified the performance of the redundancy resolution and obstacle avoidance scheme in executing the following tasks: position tracking, orientation tracking, static and moving obstacle collision avoidance, joint-limiting, and self-collision avoidance. In each of these scenarios one or multiple features were active at different instants of execution.

Despite the geometrical complexity of REDIESTRO, the arm is entirely modeled by decomposition of the links and the attached actuators into sub-links modeled by simple volume primitives. Moreover, because of the complex and unusual shape of REDIESTRO, it is believed that adapting the algorithms to other industrial and research manipulators can only be simpler.

A comparison between different methodologies for force and compliant motion control indicated that the hybrid impedance control approach is at present the most suitable scheme for compliant motion and force control. The outcome of this survey also showed that there exists no unique framework for compliant motion and force control of redundant manipulators which enjoys the following advantages:

1- Takes full advantage of redundant degrees of freedom

- by incorporating different additional tasks without modifying the scheme and the control law.

- using the redundant degrees-of-freedom to fulfill dynamic tasks such as multiple-point force control.

- using task priority and singularity robustness formulation to cop with kinematic and artificial singularities.

2- Compatibility for execution of both force and compliant motion tasks

- ensuring accurate force regulation

- achieving stable motion control in the presence of disturbance forces

3- Robustness

- with respect to higher-order unmodeled dynamics (i.e., joint flexibilities), uncertainties in manipulator dynamic parameters, and friction model and parameters.

- with respect to poor knowledge of the environment and kinematic errors

4- Adaptive implementation

- allowing for easy incorporation of adaptation in the case of manipulators for which estimates of the dynamic parameters are not available.

The Augmented Hybrid Impedance Control (AHIC) scheme presented in this monograph enjoys the aforementioned desirable characteristics. The feasibility of the scheme was evaluated by computer simulation on a 3-DOF planar arm. The most useful additional tasks – joint limit avoidance, static and moving object avoidance, self collision avoidance, and posture optimization – were incorporated into the AHIC scheme. The simulation performed for multiple point contact force control indicated one of the major characteristics of the AHIC scheme which distinguishes it from similar schemes. The additional task can not only be position-controlled but can also be included into the force-controlled subspace. This increases the capability of the redundancy resolution scheme.

The simulations on the 3-DOF planar arm showed that a simple extension of the redundancy resolution scheme at the acceleration level using the solution which minimizes the norm of the joint acceleration vector has the shortcoming that it cannot control the null-space components of joint velocities and may result in "internal instability". A modified AHIC scheme was presented that addresses this undesirable self motion problem.

The extension of the AHIC scheme to the 3D workspace of a 7-DOF manipulator (REDIESTRO) was described in Chapter 5. The complexity of the scheme required an algorithm development procedure which incorporates a high level of optimization. At the same time, the following problems were addressed in extending the modules to a 3-D workspace:

- An AHIC module for orientation and torque

- Uncontrolled self-motion due to resolving redundancy at the acceleration level for the AHIC scheme (the solution proposed in Chapter 4 is computationally expensive).

- Robustness issues with respect to higher-order unmodeled dynamics (joint flexibilities), uncertainties in manipulator dynamic parameters, and friction model and parameters.

A realistic dynamic simulation environment enables one to study issues such as performance degradation due to imprecise dynamic modeling and uncontrolled self-motion. The least-square solutions for redundancy resolution at the acceleration level were modified by adding a velocity dependent term to the cost function. This modification successfully controlled the self-motion of the manipulator.

It was demonstrated by simulation that the force tracking performance of the methods based solely on inverse dynamics degrade in the presence of uncertainty in the manipulator's dynamic parameters and unmodeled dynamics. This is especially true for a manipulator equipped with harmonic drive transmissions, which introduce a high level of joint flexibility and frictional effects (as in the case of REDIESTRO). The AHIC control scheme was modified by incorporating an "error reference controller". This modification successfully copes with model uncertainties in the model-based part of the controller, and even friction compensation is not required. The modified AHIC scheme increases the applicability of this type of control to a large class of industrial and research arms.

Chapter 6 described the experimental work carried out to evaluate the performance of the AHIC scheme in compliant motion and force control of REDIESTRO. Considering the complexity and the computation involved in force and compliant motion control of a 7-DOF redundant manipulator, the implementation of the real-time controller from both hardware and software points of view, by itself represents a challenge. It should be noted that there are few cases to date where experimental results for force and compliant motion control of a 7-DOF manipulator have been reported. Moreover, implementation of the AHIC scheme for REDIESTRO introduces additional challenges:

- The REDIESTRO arm is equipped with harmonic drive transmissions which introduce a high level of joint flexibility and make accurate control of contact force more difficult.

- The friction model that is most commonly used is that of load independent Coulomb and viscous friction. This model is especially inadequate for a robot with harmonic drive transmissions which have high friction - Experimental results on REDIESTRO show that in some configurations the friction corresponds to up to 30% of the applied torque. Also, other experimental studies have shown that frictional forces in harmonic drives are very nonlinear and load dependent.

- Performing tasks such as peg-in-the-hole involves accurate positioning. This requires a well-calibrated arm. Considering the fact that REDIESTRO has not been accurately calibrated, successful execution of the peg-in-the-hole strawman task by REDIESTRO demonstrates a high level of robustness of the scheme presented in this monograph.

Two strawman tasks: Surface cleaning and peg-in-the-hole, were selected. The selection was based on the ability to evaluate force and position tracking and also robustness with respect to knowledge of the environment and kinematic errors. The results for the surface-cleaning strawman task indicate that when there is no motion in the x and y directions, the force tracking is almost perfect (0.04 N steady-state error). When the eraser is moving on the surface, it was observed that because of unmodeled flexibility in the joints, the force tracking may degrade drastically. However, by an appropriate selection of the controller's cutoff frequency in a force-controlled direction, it is possible to achieve an acceptable level of force tracking. The experiment shows that with 20N desired force, the interaction force with mean value -19.6N and standard deviation 0.6N was achieved. This demonstrates the capability of the force-controller in regulating interaction forces even in the presence of unmodeled dynamics (joint frictions and flexibilities).

Strawman task II was also successfully demonstrated. Considering the tolerances used in the peg and the holes (0.5 mm between the peg and the hole) and the fact that REDIESTRO has not been accurately calibrated, the successful peg-in-the-hole demonstration indicates the robustness of the scheme with respect to poor knowledge of the environment.

Appendix A Kinematic and Dynamic Parameters of REDIESTRO

This appendix summarizes the kinematic and dynamic parameters of REDIESTRO, a seven-dof experimental redundant manipulator. It also provides the mechanical specification of the actuators and related hardware.

Table A-1 D-H parameters of REDIESTRO

i	α_{i-1} (deg)	a(i-1) mm	b(i) mm	q(i)[a]
1	0.	0.	952.29	q(1)
2	-58.31	0.	-22.91	q(2)
3	-20.0289	231.13	36.93	q(3)
4	105.26	0.	0.	q(4)
5	60.91	398.84	-471.59	q(5)
6	59.88	0.	578.21	q(6)
7	-75.47	135.59	-145.05	q(7)
Tool	0	234.44	0	0

a. Isotropic Configuration: q = [q1, -11.01, 91.94, 113.93, -2.26, 150.25, 63.76]

Table A-2 Mass (Kg)

Link1	Link2	Link3	Link4	Link5	Link6	Link7
17.313	5.580	28.586	7.390	5.987	2.557	0.200

Table A-3 Center of gravity in local frame {i}

.	Link1	Link2	Link3	Link4	Link5	Link6	Link7
X	0.00048	0.1155	-0.0011	0.3071	0	0.0919	0.06345
Y	-0.1607	-0.0036	-0.1176	-0.0408	-0.1326	-0.0343	0
Z	-0.1186	-0.0389	-0.1539	0.0699	0.1507	-0.0882	-0.0034

Table A-4 Link inertia tensor $(Kg\ m^2)^a$

.	Link1	Link2	Link3	Link4	Link5	Link6	Link7
Ixx	0.89926	0.02573	1.6620	0.09297	0.8284	0.67522	0.004435
Iyy	0.31342	0.13223	0.7860	0.8881	0.7019	0.69288	0.005547
Izz	0.62745	0.11099	0.9387	0.8753	0.1317	0.03904	0.001136
Ixy	-2.7e-5	-0.0045	0.0001	-0.1203	0.00009	-0.00914	0.0
Iyz	0.3689	0.0012	0.1221	-0.0204	0.26852	-0.04921	0.0
Izx	-1.2e-5	-0.0404	0.0003	0.1411	0.00016	0.13028	-0.00189

a. The inertia tensor (in the frame located at the center of gravity
   with the same orientation as the local frame {i}) is defined by:

$$I_{C^A_i} = \begin{bmatrix} I_{xx} & -I_{xy} & -I_{zx} \\ -I_{xy} & I_{yy} & -I_{yz} \\ -I_{zx} & -I_{yz} & -I_{zz} \end{bmatrix}$$

Table A-5 Motor assembly parameters

	1	2	3	4	5	6	7
Encoder resolution[a] (pulse/rev.)	200	360	360	360	1000	1000	1000
Gear ratio	200	260	260	260	160	160	110
Torque constant[b](Nm/A)	40	55	55	55	32	32	5.76
Maximum input current (A)	4.9	8.1	8.1	8.1	3.1	3.1	4.1
Actuator moment of inertia[c] (Kg m^2)	10.1	57.4	57.4	57.4	2.43	2.43	0.11
Coulomb friction(N.m)	19.2	47.3	47.3	47.3	10.24	10.24	0.92
Stiction (N.m)	15.36	25.84	25.84	25.84	8.2	8.2	0.74
Viscous coefficient (Nm.s/Rad)	0.14	0.34	0.34	0.34	0.09	0.09	0.02

a. The encoder resolution is four times greater (4*Encoder resolution) if the quadrature feature is used
b. Specified at the output shaft
c. Specified at the output shaft

Table A-6 Motor assembly interface specifications

		1	2	3	4	5	6	7
Encoders	Interface card resolution (bit)	32	32	32	32	32	32	32
Motors	Current amplifier gain (A/V)	1.2	1.2	1.2	1.2	1.2	1.2	1.2
	Current amplifier Max. Current[a] (A)	12	12	12	12	12	12	12
	DAC (bits)	12	12	12	12	12	12	12
	DAC: Max. Output (V)	10	10	10	10	10	10	10
Force sensor (JR3)	Receiver card (bits)	14	14	14	14	14	14	14
	Max. Force	fx, fy = 200N, fz = 400 N; mx,my, mz = 12.5 Nm,						

a. This can be adjusted by changing a resistor in the hardware. For the experimental study, it was set to the maximum allowable current for each motor.

Appendix B   Trajectory Generation (Special Consideration for Orientation)

The desired orientation at the end of each segment is specified by the user in a pre-programed task file. This orientation is specified in the form of X-Y-Z Fixed Angles [16]. In this representation, the orientation is specified by a 3 dimensional vector $[\gamma, \beta, \alpha]$ which can be converted to a Direction Cosine Matrix (DCM) representation as follows:

$$R_{XYZ}(\gamma, \beta, \alpha) = R_Z(\alpha)R_Y(\beta)R_X(\gamma)$$

$$R_{XYZ}(\gamma, \beta, \alpha) = \begin{bmatrix} c\alpha c\beta & c\alpha s\beta s\gamma - s\alpha c\gamma & c\alpha s\beta c\gamma + s\alpha s\gamma \\ s\alpha c\beta & s\alpha s\beta s\gamma + c\alpha c\gamma & s\alpha s\beta c\gamma - c\alpha s\gamma \\ -s\beta & c\beta s\gamma & c\beta c\gamma \end{bmatrix} \quad (B.1)$$

Let us assume that the initial orientation $[\gamma_i, \beta_i, \alpha_i]$ and final orientation $[\gamma_f, \beta_f, \alpha_f]$ are specified. Then the equivalent angle-axis representation is calculated based on the method given in [16]. Having calculated the initial orientation vector $[K_{x_i}, K_{Y_i}, K_{Z_i}]$ and the final vector $[K_{x_f}, K_{Y_f}, K_{Z_f}]$ in the angle-axis form, the fifth-order trajectory generator can be used to find the desired orientation vector $K(t)$.

It should be noted that the first and second derivatives $(\dot{K}(t), \ddot{K}(t))$ of the desired orientation vector are not the angular velocity ω and acceleration Ω respectively. The open-loop simulation (see Figure B.1) shows the robot's orientation $K_{Robot}(t)$. It does not follow the desired orientation $K(t)$. The desired angular velocity and acceleration can be calculated as follows:

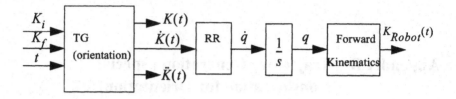

Figure B.1 Block diagram of the open-loop simulation for orientation TG.

$$[\omega, \dot{\omega}] = f(K(t), \dot{K}(t), \ddot{K}(t)) \tag{B.2}$$

A derivation of the above function is given below. The calculation of the angle-axis formulation from the DCM representation is as follows:

$$K(t) = [K_x, K_y, K_z]^T = k(t)\theta(t) \tag{B.3}$$

where $\theta(t) = \|K(t)\|$ and $k(t) = \dfrac{K(t)}{\theta(t)}$.

$$
R = \begin{bmatrix}
k_x{}^2 \upsilon\theta + c\theta & k_x k_y \upsilon\theta - k_z s\theta & k_x k_z \upsilon\theta + k_y s\theta \\
k_x k_y \upsilon\theta + k_z s\theta & k_y{}^2 \upsilon\theta + c\theta & k_y k_z \upsilon\theta - k_x s\theta \\
k_x k_z \upsilon\theta - k_y s\theta & k_y k_z \upsilon\theta + k_x s\theta & k_z{}^2 \upsilon\theta + c\theta
\end{bmatrix}
$$

$$
= \begin{vmatrix}
a_x & n_x & s_x \\
a_y & n_y & s_y \\
a_z & n_z & s_z
\end{vmatrix} \tag{B.4}
$$

where $\upsilon\theta = 1 - c\theta$. This yields

$$tr(R) = 2c\theta + 1 \quad where \quad tr(R) = a_x + n_y + s_z \tag{B.5}$$

$$k = \frac{vect(R)}{s\theta} \quad where \quad vect(R) = \frac{1}{2}\begin{bmatrix} n_z - s_y \\ s_x - a_z \\ a_y - n_x \end{bmatrix} \tag{B.6}$$

Now, we differentiate with respect to time to get

$$\dot{K}(t) = \dot{k}(t)\theta(t) + k(t)\dot{\theta}(t) \tag{B.7}$$

We need to find $\dot{k}, \dot{\theta}$ as a linear function of K, ω. To do this, we note that

$$\frac{d}{dt}(vect(R)) = vect(\dot{R}) \tag{B.8}$$

and

$$\dot{R}R^{-1} = \Omega = \begin{bmatrix} 0 & -\omega_z & \omega_y \\ \omega_z & 0 & -\omega_x \\ -\omega_y & \omega_x & 0 \end{bmatrix} \tag{B.9}$$

So that

$$vect(\dot{R}) = vect(\Omega R) = \frac{1}{2}X\omega \text{ where } X = tr(R)I - R \tag{B.10}$$

and

$$tr(\dot{R}) = Tr(\Omega R) = -2s\theta k^T \omega \tag{B.11}$$

Now (B.6) yields

$$\dot{k} = \frac{(tr(R)I - R)\omega}{2s\theta} - \frac{c\theta k\dot{\theta}}{s\theta} \tag{B.12}$$

Differentiating (B.5) with respect to time results in

$$\dot{\theta} = \frac{tr(\dot{R})}{-2s\theta} \tag{B.13}$$

Substituting (B.11) into (B.13) yields

$$\dot{\theta} = k^T \omega \tag{B.14}$$

From equations (B.12) and (B.14), we get

$$\omega = 2s\theta N^{-1}\dot{k} \tag{B.15}$$

where

$$N = \theta M + 2s\theta k k^T \quad and \quad M = tr(R)I - R - 2c\theta k k^T \tag{B.16}$$

Substituting in (B.7) from (B.12) and (B.14) results in

$$2s\theta\dot{K} = M\omega\theta + 2s\theta kk^T\omega = F\omega \qquad (B.17)$$

where

$$F = M\theta + 2s\theta kk^T \qquad (B.18)$$

Differentiating (B.17) yields

$$2c\theta\dot{K} + 2s\theta\ddot{K} = \dot{F}\omega + F\dot{\omega} \qquad (B.19)$$

$$\dot{\omega} = F^{-1}(2c\theta\dot{K} + 2s\theta\ddot{K} - \dot{F}\omega) \qquad (B.20)$$

Now, we need to find $\dot{F}$

$$\dot{F} = \dot{M}\theta + M\dot{\theta} + 2c\theta\dot{\theta}kk^T + 2s\theta(\dot{k}k^T + k\dot{k}^T) \qquad (B.21)$$

where

$$\dot{M} = -2s\theta k^T\omega I - \Omega R + 2s\theta k^T\omega kk^T - 2c\theta(\dot{k}k^T + k\dot{k}^T) \qquad (B.22)$$

The optimized C code for this function is produced by the symbolic optimization routine provided by the RDM software [78].

References

[1] R. Anderson, and M.W. Spong, "Hybrid impedance control of robotic manipulators", *Proc. IEEE Int. Conf. on Robotics and Automation*, pp. 1073-1080, 1987.

[2] N. Adachi, Z.X. Peng, and S. Nakajima, "Compliant motion control of redundant manipulators", *IEEE/RSJ Workshop on Intell. Rob. Sys.*, pp. 137-141, 1991.

[3] J. Angeles, F. Ranjbaran, and R.V. Patel, "On the design of the kinematic structure of seven-axes redundant manipulators for maximum conditioning", *Proc. IEEE International Conf. Robotics and Automation*, pp.494-499, 1992.

[4] J. Angeles, "The Application of Dual Algebra to Kinematic Analysis", in Angeles, J. and Zakhariev, E. (editors), *Computational Methods in Mechanical Systems,* Springer-Verlag, Heidelberg, vol. 161, pp. 3-31, 1998.

[5] J. Angeles, *Fundamentals of Robotic Mechanical Systems: Theory, Methods, and Algorithms,* 2nd Edition, Springer-Verlag, New York, 2002.

[6] J. Baillieul, "Avoiding Obstacles and resolving kinematic redundancy", *Proc. IEEE Int. Conf. on Robotics and Automation*, pp. 1698-1704, 1986.

[7] S. Borner, and R. B. Kelley, "A novel representation for planning 3-D collision free paths", *IEEE Transaction on Syst., Man, and Cybernetics*, vol. 20, no. 6, pp. 1337-1351, 1990.

[8] P. Borrel, "Contribution a la modelisation geometrique des robots manipulateurs : Application a la conception assistee par l'ordinateur", *These d'Etat*, USTL, Montpellier, France, July 1986.

[9]   I.J. Bryson, *Software Architecture and Associated Design and Implementation Issues for Multiple-Robot Simulation and Vizualization*, M.E.Sc. Thesis, University of Western Ontario, London, Ontario, 2000.

[10]   I.J. Bryson and R.V. Patel, "A modular software architecture for robotic simulation and visualization", *31st Int. Symposium on Robotics (ISR2000)*, Montreal, Canada, May 14-17, 2000.

[11]   Y. Bu and S. Cameron, "Active motion planning and collision avoidance for redundant manipulators", *1997 IEEE Int. Symposium on Assembly and Task Planning*, pp. 13-18, Aug. 1997.

[12]   B.W. Char, et al., *Maple V Language Reference Manual*, Springer-Verlag, New York, 1991

[13]   S.L. Chiu, "Task compatibility of manipulator postures", *Int. Journal of Robotics Research*, vol. 7, no. 5, pp 13-21, Oct. 1988.

[14]   R. Colbaugh, H. Seraji, and K. Glass, "Obstacle avoidance of redundant robots using configuration control", *Int. Journal of Robotics Research*, vol. 6, pp. 721-744, 1989.

[15]   Colombina, F. Didot, G. Magnani, and A. Rusconi, "External servicing testbed for automation and robotics", *IEEE Robotics & Automation Magazine*, Mar. 1996, pp. 13-23

[16]   J.J. Craig, *Introduction to Robotics: Mechanics and Control*, 2nd Edition, Addison Wesley, 1995.

[17]   D. Dawson and Z. Qu, "Comments on impedance control with adaptation for robotic manipulators", *IEEE Trans. on Robotics and Automation*, vol 7, no. 6, Dec. 1991.

[18]   A. De Luca, "Zero Dynamics in Robotic Systems", in *Nonlinear Synthesis*, C. I. Byrnes and A. Kurzhanski (Eds.), Progress in Systems and Control Series, Birkhauser, Boston, MA, 1991.

[19]   R.V. Dubey, J.A. Euler, and S.M. Babock, " An efficient gradient projection optimization scheme for a seven-degree-of-freedom redundant robot with spherical wrist", *Proc. IEEE Int. Conf. on Robotics and Automation*, pp. 28-36, Philadelphia, PA, 1988.

[20]   J. Duffy, "The fallacy of modern hybrid control theory that is based on "orthogonal complements" of twist and wrench spaces", *Journal of Robotic Systems*, vol. 7, no. 2, pp. 139-144, 1990.

[21]   O. Egeland, "Task-space tracking with redundant manipulators", *IEEE Journal of Robotics and Automation*, vol. 3, pp. 471-475, 1987.

[22]   Q.J. Ge, "An inverse design algorithm for a G2 interpolating spline motion", in *Advances in Robot Kinematics and Computational Geometry*, J. Lenarcic and B. Ravani (eds.), Kluwer Academic Publishers, Norwell, MA, pp. 81-90, 1994

[23]   M.W. Gertz, J. Kim, and P. Khosla, "Exploiting redundancy to reduce impact force", *IEEE/RSJ Workshop on Intell. Rob. Sys*, pp. 179-184, 1991.

[24]   K. Glass, R. Colbaugh, D. Lim, and H. Seraji, "Real-time Collision avoidancefor redundant manipulators", *IEEE Transaction on Robotics and Automation*, pp. 448-457, vol 11. no. 10, 1995.

[25]   G.H. Golub and C.F. Van Loan, *Matrix Computations,* 2nd ed., John Hopkins Univ. Press, Baltimore, 1989.

[26]   M.A. Gonzalez-Palacios, J. Angeles and F. Ranjbaran, "The kinematic synthesis of serial manipulators with a prescribed Jacobian", *Proc. IEEE Int. Conf. Robotics Automat.*, Atlanta, Georgia, 1993, vol. 1, pp 450-455.

[27]   Y. Han, L. Liu, R. Lingarkar, N. Sinha, and M. Elbestawi, "Adaptive Control of Constrained Robotic manipulators", *Int. Journal of Robotics Research*, vol. 7, no. 2, pp. 50-56, 1992.

[28]   T. Hasegawa and H. Terasaki, "Collision avoidance: Divide-and-conquer approach by space characterization and intermediate goals", *IEEE Transaction on Syst., Man, and Cybernetics,* vol. 18, no. 3, pp. 337-347, 1988.

[29]   H. Hattori and K. Ohnishi, "A realization of compliant motion by decentralized control in redundant manipulators", *Proc. IEEE/ ASME Int. Conf. on Advanced Intelligent Mechatronics,* Como, Italy, pp. 799-803, 2001.

[30]   N. Hogan, "Impedance control: An approach to manipulation", *ASME Journal of Dynamic Systems, Measurment, and Control,* vol. 107, pp. 8-15, 1985.

[31]   J.M. Hollerbach, and K.C. Suh, "Redundancy resolution for manipulators through torque optimization", *Int. Journal of Robotics Research*, vol. 3, pp. 308-316, 1987.

[32]   P. Hsu, J. Hauser, and S. Sastry, "Dynamic Control of Redundant Manipulators", *Journal of Robotic Systems*, vol. 6, pp. 133-148,1989.

[33]   D.G. Hunter, "An overview of the Space Station Special Purpose Dexterous Manipulator", *National Research Council Canada*, NRCC no. 28817, Issue A, 7 April 1988

[34]   H. Kazerooni., T.B. Sheridan, and P.K. Houpt, "Robust compliant motion for manipulators: Part I: The fundamental concepts of compliant motion", *IEEE Trans. on Robotics and Automation*, vol. 2, no. 2, pp. 83-92, 1986.

[35]   K. Kazerounian and Z. Wang, "Global versus local optimization in redundancy resolution of robotics manipulators", *Int. Journal of Robotics Research*, vol. 7. no. 5, pp.3-12, 1988.

[36]   P. Khosla and R.V. Volpe, "Superquadric artificial potentials for an obstacle avoidance approach", *Proc. IEEE Int. Conf. on Robotics and Automation*, pp. 1778-1784, 1988.

[37]   C.A. Klein and C.H. Hung, "Review of pseudoinverse control for use with kinematically redundant manipulators", *IEEE Trans. on Systems, Man, and Cybernetics*, vol. 13, pp. 245-250, 1983.

[38]   C.A. Klein, "Use of redundancy in design of robotic systems", *Proc. 2nd Int. Symp. Robotic Res.*, Kyoto, Japan, 1984.

[39]   Z. Lin, R.V. Patel, and C.A. Balafoutis, "Augmented impedance control: An approach to impact reduction for kinematically redundant manipulators", *Journal of Robotic Systems*, vol. 12, pp. 301-313, 1995.

[40]   G.J. Liu and A.A. Goldenberg, "Robust hybrid impedance control of robot manipulators", *Proc. IEEE Int. Conf. on Robotics and Automation*, pp. 287-292, 1991.

[41]   W.S. Lu, and Q.H. Meng, "Impedance control with adaptation for robotic manipulators", *IEEE Trans. on Robotics and Automation*, vol 7, no. 3, June 1991.

[42]   J.Y.S. Luh, M.W. Walker, and R.P.C. Paul, "Resolved-acceleration of mechanical manipulators", *IEEE Transaction on Automatic Control*, vol. AC-25, no. 3, pp. 468-474, June 1980.

[43] A.A. Maciejewski and C.A. Klein, "The singular value decomposition: Computation and application to robotics", *Int. Journal of Robotics Research*, vol. 8, no. 6, Dec. 1989.

[44] *Matlab External Interface Guide for UNIX Workstation*, The Math-Works Inc., 1992.

[45] N.H. McClamroch and D. Wang, "Feedback stablization and tracking in constrained robots", *IEEE Trans. on Automatic Control*, vol. 33, no. 5, pp. 419-426, 1988.

[46] J.K. Mills, "Hybrid Control: A constrained motion perspective", *Journal of Robotic Systems*, vol. 8, N0. 2, pp. 135-158, 1991.

[47] Y. Nakamura and H. Hanafusa, "Inverse kinematic solutions with singularity robustness for manipulator control", *ASME Journal of Dynamic Systems, Measurment, and Control*, vol. 108, pp.163-171, 1986.

[48] Y. Nakamura and H. Hanafusa, "Optimal redundancy control of robot manipulators", *Int. Journal of Robotics Research*, vol. 6, no. 1, pp. 32-42, 1987.

[49] K.S. Narendra and A.M. Annaswamy, *Stable Adaptive Systems*, Prentice Hall, Englewood cliffs, NJ, 1989.

[50] B. Nemec and L. Zlajpah, "Force control of redundant robots in unstructured environments", *IEEE Trans. on Industrial Electronics*, vol. 49, no. 1, pp. 233-240, 2002.

[51] W.S. Newman and M.E. Dohring, "Augmented impedance control: An approach to compliant control of kinematically redundant manipulators", *Proc. IEEE International Conf. Robotics and Automation*, pp. 30-35, 1991.

[52] G. Niemeyer and J.J. Slotine, "Performance in adaptive manipulator control", *Int. Journal of Robotics Research*, vol. 10, no. 2, April 1991.

[53] Y. Oh, W.K. Chung, Y. Youm, and I.H. Suh, "Motion/force decomposition of redundant manipulators and its application to hybrid impedance control", *Proc. IEEE Int. Conf. on Robotics and Automation*, Leuven, Belgium, pp. 1441-1446, 1998.

[54] R. Ortega and M. Spong, "Adaptive motion control of rigid robots: A tutorial." In *Proc. IEEE conf. on Decision and Control.*, Austin, Texas, 1988.

[55]   R.P.C. Paul, *Robot Manipulators,* MIT Press, Cambridge, MA, pp. 28-35, 1981.

[56]   M. Raibert and J.J. Craig, "Hybrid position-force control of manipulators", *ASME Journal of Dynamic Systems, Measurment, and Control,* vol. 102, pp. 126-133, 1981.

[57]   F. Ranjbaran, J. Angeles, M.A. Gonzalez-Palacios, and R.V. Patel , "The mechanical design of a seven-axes manipulator with kinematic isotropy", *Journal of Robotics and Intelligent Systems,* vol. 13, pp. 1-21, 1995.

[58]   *REACT in IRIX 5.3,* Technical Report, Silicon Graphics Inc., Dec. 1994.

[59]   N . Sadegh, and R. Horowitz, "Stability analysis of an adaptive controller for robotic manipulator", *in Proc. IEEE Int. Conf. Robotics and Automation.*

[60]   K.J. Salisbury, "Active stiffness control of manipulators in Cartesian coordinates", *Proc. IEEE Int. Conf. on Robotics and Automation,* pp. 95-100, 1980.

[61]   L. Sciavicco and B. Siciliano, "A solution algorithm to the inverse kinematic problem of redundant manipulators", *IEEE Journal of Robotics and Automation,* vol. 4, pp. 403-410, 1988.

[62]   L. Sciavicco and B. Siciliano, "An algorithm for reachable workspace for 2R and 3R planar pair mechanical arms", *Proc. IEEE Int. Conf. Robotics and Automation,* vol. 1, pp 628-629, Philadelphia, PA,1988.

[63]   H. Seraji, "Configuration control of redundant manipulators: Theory and implementation", *IEEE Transactions on Robotics and Automation,* vol. 5, pp. 472-490, 1989.

[64]   H. Seraji and R. Colbaugh, "Improved Configuration Control for redundant robots", *Journal of Robotic Systems,* vol. 7, no. 6, pp. 897-928, 1990.

[65]   H. Seraji and R. Colbaugh, "Singularity-robustness and task prioritization in configuration control of redundant robots", *29th IEEE Conf. on Decision and Control,* pp. 3089-3095,1990.

[66]   H. Seraji, "Task options for redundancy resolution using configuration control", *30th IEEE Conf. on Decision and Control,* pp. 2793-2798, 1991.

[67]   H. Seraji, D. Lim, and R. Steele, "Experiments in contact control", *Journal of Robotic Systems*, vol. 13, no. 2, pp. 53 - 73, 1996.

[68]   H. Seraji, R. Steele, and R. Ivlev, "Sensor-based collision avoidance: Theory and experiments", *Journal of Robotic Systems,* vol. 13, no. 9, pp. 571-586, 1996.

[69]   H. Seraji and R. Steele, "Nonlinear contact control for space station dexterous arms", *Proc. IEEE Int. Conf. on Robotics and Automation*, Leuven, Belgium, pp. 899-906, 1998.

[70]   H. Seraji and B. Bon, "Real-time collision avoidance for position-controlled manipulators", *IEEE Trans. on Robot. and Automat.*, vol. 15, no. 4, pp. 670-677, 1999.

[71]   F. Shadpey, C. Tessier, R.V. Patel, and A. Robins, "A trajectory planning and obstacle avoidance system for kinematically redundant manipulators", *CASI Conference on Astronautics*, Ottawa, Nov. 1994.

[72]   F. Shadpey, R.V. Patel, C. Balafoutis, and C. Tessier, " Compliant Motion Control and Redundancy Resolution for Kinematically Redundant Manipulators", *American Control Conference*, Seattle, WA, June 1995.

[73]   F. Shadpey and R.V. Patel, "Compliant motion control with self-motion Stabilization for kinematically redundant manipulators", *Third IASTED Int. Conf. on Robotics and Manufacturing*, Cancun, Mexico, June 1995.

[74]   F. Shadpey and R.V. Patel, "Adaptive Compliant Motion Control of Kinematically Redundant Manipulators", *IEEE Conf. on Decision and Control*, Dec. 1995.

[75]   F. Shadpey, C. Tessier, R.V. Patel, B. Langlois, and A. Robins, "A trajectory planning and object avoidance system for kinematically redundant manipulators: An experimental evaluation", *AAS/AIAA American Astrodynamics Conference*, Aug. 1995, Halifax, Canada.

[76]   F. Shadpey, F. Ranjbaran, R.V. Patel, and J. Angeles, "A compact cylinder-cylinder collision avoidance scheme for redundant manipulators", *Sixth Int. Symp. on Robotics and Manufacturing (ISRAM)*, Montpellier, France, May, 1996.

[77]   F. Shadpey, M. Noorhosseini, I. Bryson, and R.V. Patel, "An inte-
        grated robotic development environment for task planning and
        obstacle Avoidance",   *Third ASME Conf. on Eng. System Design
        & Analysis*, Montpellier, France, July 1996.

[78]   F. Shadpey and R.V. Patel, "Robot Dynamic Modelling (RDM)
        Software: User's Guide", Concordia University, Montreal, Canada,
        Feb. 1997.

[79]   K. Shoemake, "Animating rotation with quaternion curves", *ACM
        Siggraph,* vol. 19, no. 3, pp. 245-254, 1985.

[80]   P.R. Sinha and A.A. Goldenberg, "A unified theory for hybrid con-
        trol of manipulators", *Proc. IFAC 12th World Congress*, Sydney,
        Australia,1993.

[81]   J.J. Slotine and W. Li, "On the Adaptive Control of Robot Manipu-
        lators", *Int. Journal of Robotics Research*, vol. 6, no. 3, pp. 49-59,
        1987.

[82]   J.J. Slotine, and W. Li, *Applied Nonlinear Control,* Prentice Hall,
        Englewood cliffs, NJ, 1991.

[83]   D.B. Stewart, R.A. Volpe, and P.K. Khosla, "Integration of Real-
        Time Software Module for Reconfigurable Sensor-Based Control
        Systems", *Proc. IEEE/RSJ International Conference on Intelligent
        Robots and Systems (IROS '92)*, Raleigh, North Carolina, pp. 325-
        332, 1992.

[84]   K.C. Suh and J.M. Hollerbach, "Local versus global torque optimi-
        zation of redundant manipulators", *Proc. IEEE Int. Conf. on Robot-
        ics and Automation*, pp. 619-624, 1987.

[85]   M. Tandirci, J. Angeles, and F. Ranjbaran, "The characteristic point
        and characteristic length of robotic manipulators", *22nd ASME
        Biennial Mechanics Conference,* Sep. 13-16, Scottsdale, AZ, vol.
        45, pp. 203-208, 1992.

[86]   C.-P. Teng and J. Angeles, "A sequential-quadratic programming
        algorithm using orthogonal decomposition with Gerschgorin stabili-
        zation", *ASME J. Mechanical Design*, vol. 123, Dec. 2001, pp. 501-
        509.

[87]   C. Tessier et al., *Trajectory Planning and Object Avoidance (STEAR
        5) - Phase II*, Final Report, vol. 1, DSS Canada, Contract no. 9F006-
        2-0107/01-SW , 1995.

[88] T.D. Tuttle and W.P Seering, "A nonlinear model of a harmonic drive gear transmission", *IEEE Trans. Rob. and Aut.*, vol. 12, no., 3, June 1996.

[89] R.S. Varga, *Matrix Iterative Analysis*, Springer-Verlag, New York, 2000.

[90] G.R., Veldkamp, "On the use of dual numbers, vectors and matrices in instantaneous, spatial kinematics", *Mechanism and Machine Theory*, vol. 11, pp. 141-156, 1976.

[91] I.D. Walker, "The use of kinematic redundancy in reducing impact and contact effects in manipulation", *Proc. IEEE International Conf. Robotics and Automation*, pp. 434-439, 1990.

[92] C.W. Wampler, "Manipulator inverse kinematic solution based on vector formulation and damped least-squares methods", *IEEE Trans. on Systems, Man, and Cybernetics*, vol. 16, no. 1. pp. 93-101, 1986.

[93] D.S. Watkins, *Fundamentals of Matrix Computations*, 2nd Edition, John Wiley & Sons, New York, 2002.

[94] D.E. Whitney, "Historical perspective and state of the art in robot force control", *Int. Journal of Robotics Research*, vol. 6, no. 1, Dec. 1987.

[95] A.T. Yang and F. Freudenstein "Application of dual-number quaternion algebra to the analysis of spatial mechanisms", *Trans. ASME J. Appl. Mech.*, pp. 300-308, 1964.

[96] T. Yoshikawa, "Dynamic hybrid position/force control of robot manipulators", *IEEE Journal of Robotics and Automation*, vol. 3, no. 5, pp. 386-392, 1987.

[97] T. Yoshikawa, "Analysis and control of robot manipulators with redundancy", *Rob. Res., 1st Int. Symp.*, MIT Press, pp. 735-747, 1984.

Index

Lecture Notes in Control and Information Sciences

Edited by M. Thoma and M. Morari

Further volumes of this series can be found on our homepage:
springeronline.com

Vol. 288: Taware, A. and Tao, G.
Control of Sandwich Nonlinear Systems
393 p. 2003 [3-540-44115-8]

Vol. 287: Mahmoud, M.M.; Jiang, J.; Zhang, Y.
Active Fault Tolerant Control Systems
239 p. 2003 [3-540-00318-5]

Vol. 286: Rantzer, A. and Byrnes C.I. (Eds)
Directions in Mathematical Systems
Theory and Optimization
399 p. 2003 [3-540-00065-8]

Vol. 285: Wang, Q.-G.
Decoupling Control
373 p. 2003 [3-540-44128-X]

Vol. 284: Johansson, M.
Piecewise Linear Control Systems
216 p. 2003 [3-540-44124-7]

Vol. 283: Fielding, Ch. et al. (Eds)
Advanced Techniques for Clearance of
Flight Control Laws
480 p. 2003 [3-540-44054-2]

Vol. 282: Schröder, J.
Modelling, State Observation and
Diagnosis of Quantised Systems
368 p. 2003 [3-540-44075-5]

Vol. 281: Zinober A.; Owens D. (Eds)
Nonlinear and Adaptive Control
416 p. 2002 [3-540-43240-X]

Vol. 280: Pasik-Duncan, B. (Ed)
Stochastic Theory and Control
564 p. 2002 [3-540-43777-0]

Vol. 279: Engell, S.; Frehse, G.; Schnieder, E. (Eds)
Modelling, Analysis, and Design of Hybrid Systems
516 p. 2002 [3-540-43812-2]

Vol. 278: Chunling D. and Lihua X. (Eds)
H_∞ Control and Filtering of
Two-dimensional Systems
161 p. 2002 [3-540-43329-5]

Vol. 277: Sasane, A.
Hankel Norm Approximation
for Infinite-Dimensional Systems
150 p. 2002 [3-540-43327-9]

Vol. 276: Bubnicki, Z.
Uncertain Logics, Variables and Systems
142 p. 2002 [3-540-43235-3]

Vol. 275: Ishii, H.; Francis, B.A.
Limited Data Rate in Control Systems with Networks
171 p. 2002 [3-540-43237-X]

Vol. 274: Yu, X.; Xu, J.-X. (Eds)
Variable Structure Systems:
Towards the 21st Century
420 p. 2002 [3-540-42965-4]

Vol. 273: Colonius, F.; Grüne, L. (Eds)
Dynamics, Bifurcations, and Control
312 p. 2002 [3-540-42560-9]

Vol. 272: Yang, T.
Impulsive Control Theory
363 p. 2001 [3-540-42296-X]

Vol. 271: Rus, D.; Singh, S.
Experimental Robotics VII
585 p. 2001 [3-540-42104-1]

Vol. 270: Nicosia, S. et al.
RAMSETE
294 p. 2001 [3-540-42090-8]

Vol. 269: Niculescu, S.-I.
Delay Effects on Stability
400 p. 2001 [1-85233-291-316]

Vol. 268: Moheimani, S.O.R. (Ed)
Perspectives in Robust Control
390 p. 2001 [1-85233-452-5]

Vol. 267: Bacciotti, A.; Rosier, L.
Liapunov Functions and Stability in Control Theory
224 p. 2001 [1-85233-419-3]

Vol. 266: Stramigioli, S.
Modeling and IPC Control of Interactive Mechanical
Systems – A Coordinate-free Approach
296 p. 2001 [1-85233-395-2]

Vol. 265: Ichikawa, A.; Katayama, H.
Linear Time Varying Systems and Sampled-data Systems
376 p. 2001 [1-85233-439-8]

Vol. 264: Baños, A.; Lamnabhi-Lagarrigue, F.;
Montoya, F.J
Advances in the Control of Nonlinear Systems
344 p. 2001 [1-85233-378-2]

Vol. 263: Galkowski, K.
State-space Realization of Linear 2-D Systems with
Extensions to the General nD (n>2) Case
248 p. 2001 [1-85233-410-X]

Vol. 262: Dixon, W.; Dawson, D.M.; Zergeroglu, E.;
Behal, A.
Nonlinear Control of Wheeled Mobile Robots
216 p. 2001 [1-85233-414-2]

Vol. 261: Talebi, H.A.; Patel, R.V.; Khorasani, K.
Control of Flexible-link Manipulators
Using Neural Networks
168 p. 2001 [1-85233-409-6]

Vol. 260: Kugi, A.
Non-linear Control Based on Physical Models
192 p. 2001 [1-85233-329-4]

Vol. 259: Isidori, A.; Lamnabhi-Lagarrigue, F.;
Respondek, W. (Eds)
Nonlinear Control in the Year 2000 Volume 2
640 p. 2001 [1-85233-364-2]

Vol. 258: Isidori, A.; Lamnabhi-Lagarrigue, F.;
Respondek, W. (Eds)
Nonlinear Control in the Year 2000 Volume 1
616 p. 2001 [1-85233-363-4]

Vol. 257: Moallem, M.; Patel, R.V.; Khorasani, K.
Flexible-link Robot Manipulators
176 p. 2001 [1-85233-333-2]

Printing and Binding: Strauss GmbH, Mörlenbach